MORE WITH LESS:
WORK REORGANIZATION IN THE CANADIAN MINING INDUSTRY

The massive changes under way in capitalist commodity production include the transition from a traditional or Fordist approach to a post-Fordist one, involving practices such as employee involvement, continuous improvement, and gain-sharing. In this research monograph, Bob Russell explores the changing character of industrial relations and labour processes in two staple industries: potash and uranium mining.

Using an innovative case-analytic approach, Russell compares the managerial strategies used by five transnational firms. As indicated by his title, *More with Less*, he sees the shift toward post-Fordism as having more to do with the intensification of labour, accomplished in part through the creation of multi-tasked positions, than with worker empowerment and the transcendence of class conflict. Russell combines extensive empirical analysis with a review of contemporary writing on work relations and labour processes to provide this intensive political-economic perspective on the capital-labour relation. His meticulous research will interest scholars and professionals in Canada, the United States, Britain, Europe, and Australia.

BOB RUSSELL is a professor in the Department of Sociology, and an associate member of the Department of Women's and Gender Studies, University of Saskatchewan.

Studies in Comparative Political Economy and Public Policy
Editors: MICHAEL HOWLETT, DAVID LAYCOCK, STEPHEN MCBRIDE,
Simon Fraser University

Studies in Comparative Political Economy and Public Policy is designed to showcase innovative approaches to political economy and public policy from a comparative perspective. While originating in Canada, the series will provide attractive offerings to a wide international audience, featuring studies with local, sub-national, cross-national and international empirical bases and theoretical frameworks.

Editorial Advisory Board:
ISABEL BAKKER, *Political Science, York University*
COLIN BENNETT, *Political Science, University of Victoria*
WALLACE CLEMENT, *Sociology, Carleton University*
WILLIAM COLEMAN, *Political Science, McMaster University*
BARRY EICHENGREEN, *Economics, University of California (Berkeley)*
WYNFORD GRANT, *Political Science, University of Warwick*
JOHN HOLMES, *Geography, Queen's University*
JANE JENSON, *Political Science, Université de Montréal*
WILLIAM LAFFERTY, *Project for an Alternative Future, Oslo*
GORDON LAXER, *Sociology, University of Alberta*
RONALD MANZER, *Political Science, University of Toronto*
JOHN RAVENHILL, *Political Science, Australia National University*
PETER TAYLOR-GOOBY, *Social Work, University of Kent*
MARGARET WEIR, *Brookings Institution, Washington, D.C.*

Published to date:
1 **The Search for Political Space: Globalization, Social Movements, and the Urban Political Experience** WARREN MAGNUSSON
2 **Oil, the State, and Federalism: The Rise and Demise of Petro-Canada as a Statist Impulse** JOHN ERIK FOSSUM
3 **Defying Conventional Wisdom: Free Trade and the Rise of Popular Sector Politics in Canada** JEFFREY M. AYRES
4 **Community, State, and Market on the North Atlantic Rim: Challenges to Modernity in the Fisheries** RICHARD APOSTLE, GENE BARRETT, PETER HOLM, SVEIN JENTOFT, LEIGH MAZANY, BONNIE MCCAY, KNUT H. MIKALSEN
5 **More with Less: Work Reorganization in the Canadian Mining Industry** BOB RUSSELL

BOB RUSSELL

More with Less:
Work Reorganization in the
Canadian Mining Industry

UNIVERSITY OF TORONTO PRESS
Toronto Buffalo London

© University of Toronto Press Incorporated 1999
Toronto Buffalo London
Printed in Canada

ISBN 0-8020-4354-2 (cloth)
ISBN 0-8020-8178-9 (paper)

Printed on acid-free paper

Canadian Cataloguing in Publication Data

Russell, Bob, 1950–
  More with less : work reorganization in the Canadian mining industry

  (Studies in comparative political economy and public policy)
  Includes bibliographical references and index.
  ISBN 0-8020-4354-2 (bound)   ISBN 0-8020-8178-9 (pbk.)

  1. Industrial relations – Canada.   2. Potash mines and mining –
  Canada – Management.   3. Uranium mines and mining – Canada –
  Management.   I. Title.   II. Series.

  HD6976.M62C35 1998   331′.04223636′0971   C98-931257-7

University of Toronto Press acknowledges the financial assistance to its publishing
program of the Canada Council for the Arts and the Ontario Arts Council.

This book has been published with the help of a grant from the Humanities and Social
Sciences Federation of Canada, using funds provided by the Social Sciences and
Humanities Research Council of Canada.

To the victims of Westray

Contents

LIST OF TABLES   ix
LIST OF FIGURES   xi
ACKNOWLEDGMENTS   xiii

1: Introduction   3
'A Sign of the Times'   3
The Issues of Post-Fordism and Post-Industrialism   10
The Study: Genealogy, Design, Accident, and Politics   21

2: Market Preliminaries: Product and Labour Markets in the Potash and Uranium Industries   28
Use-Values and Exchange-Values: Markets and Prices   30
The Labour Markets   40
Summary   50

3: Corporate Cultures of Employment I: Two Traditional Firms   52
Corporate Cultures and the Management of Workers   52
PCS Inc.: Fordism and Radically Marketized Industrial Relations   54
Cameco and Radically Spatialized Industrial Relations   65

4: Corporate Cultures of Employment II: Two Post-Fordist Firms   75
Central Canada Potash: Elements of a Passive Post-Fordism   75
Agrium and the Ethic of Continuous Improvement   87
Reprise   108

5: The Labour Process   111
The Labour Processes   112
Labour Process Theory   124
Some Evidence on Skill and Related Matters   141
Conclusions   158

6: Production Politics at Five Mine Sites   162
Control and Conflict at the Mines   168
Production Politics and the Work Environment   178
Conclusions   194

7: Final Reflections   195

NOTES   201
REFERENCES   233
INDEX   243

Tables

2.1   World potash production, 1989–94   31
2.2   Canadian exports of fertilizer potash, destination by percentage share, 1993–4   32
2.3   Five largest overseas potash markets, 1990–4   34
2.4   Potash price quotations, 1972–95   35
2.5   Production of uranium in concentrates by selected major producing countries, 1988–93   38
2.6   Exports of Canadian uranium by destination, 1988–93   39
2.7   Canadian uranium export prices, 1974–94   40
2.8   Potash workforce: father's and mother's occupational background   42
2.9   Previous job histories of potash workforce   43
2.10 Uranium workforce (Key Lake), occupation by race   45
2.11 Uranium workforce: father's and mother's occupational background, Aboriginal and non-Aboriginal workers   46
2.12 Previous job histories of uranium workforce, Aboriginal and non-Aboriginal workers   47
3.1   Job classes and wage rates, PCS-Allan, 1993   61
4.1   Management rights under proposed CCP Employee Involvement Plan   77
4.2   Chronology of reforms at Central Canada Potash, late 1980s to early 1990s   79
4.3   Model for participative management at Central Canada Potash   83
4.4   Agrium occupational structure and wage structure   91
4.5   Designated roles and responsibilities in Agrium continuous improvement project   94
4.6   Agrium portrayal of old and new business paradigms   97
4.7   Summary of work organization at four firms   109

x   Tables

5.1   Job rotation   143
5.2   Use of job reassignments   144
5.3   Job expansion   145
5.4   Reported levels of staffing   146
5.5   Reported production norms   147
5.6   Skill trends   148
5.7   Skill trends by occupation at five worksites   151
5.8   Control of technology by occupation at five worksites   152
5.9   Participation in job rotation by occupation at five worksites   154
5.10 Job reassignments by occupation at five worksites   155
5.11 Job expansion by occupation at five worksites   157
5.12 Logistic regression estimates predicting skill trends at five worksites   159
6.1   Disciplinary issues   170
6.2   Grievances filed   171
6.3   Grievance issues   172
6.4   Reported levels of different types of sabotage   177
6.5   Reported levels of worker involvement   180
6.6   Willingness to become involved in joint decision making   183
6.7   Work injuries   185
6.8   Time-loss work injuries   186
6.9   Responses to unsafe work requests   188
6.10 Reported job harassment   190
6.11 Types of reported harassment, all mines   191

Figures

3.1 Lines of progression, Mine Operating Department, PCS-Allan 62
4.1 Model for the functioning of continuous improvement program at
    Agrium 93
5.1 Geological structure of potash mining 115
5.2 Key Lake Mill simplified flowsheet 123

Acknowledgments

In the most important ways this book is a collective undertaking. In short, it would not have been possible to have undertaken this project without the active cooperation and voluntarism of many individuals. Bernie Welke and Don Zacharias, permanent and acting staff representatives, respectively, of the United Steelworkers were receptive to my original intentions, and encouraged me to expand upon them, with consequences that I could not have imagined at the time. Their curiosity, openness, and trust were contagious. All possible forms of accessibility were made available through their good offices. Ken Neumann, USWA Director of District 3 (Western Provinces and NWT) was also an enthusiastic backer of the study from its initial stages onwards. Through his generosity I was invited to take part in different regional union conferences and events. The individual local union presidents at the time of the study, Gerhard Bettig, Larry Buchinski, Brian Kluchewski, Ben Medernach, Don Morden, Pete Neufeld, Ron St. Pierre, and Gord Telfer, 'adopted' me into their locals and saw that every possible resource and courtesy were proffered. Individually and collectively, their patience with what must have seemed like endless requests and clarifications knew no bounds. Subsequently, several of these individuals have read and commented upon various parts of the manuscript. Most important, each of these persons allowed me to enter their world, an industrial world that is in transition, and, I hope, to understand the changes that are taking place within it. For this I am very grateful.

For her help in setting up interviews, conducting a good portion of them, and entering the data, I am indebted to Deb Baczuk, who was with the project from inception to end. I could not have hoped to have found a more enthusiastic or helpful associate. Dale Emory and Greg Yelland were also conscientious interviewers. Thanks are also owed to Ron Templeman and Pam Smith for technical advice on survey design and file management at different points over the life of the project. Fern Hagin and Alma Roberts are responsible for the many tables in the text, their detailed accuracy and proper alignment. Norma Deaver of the Potash Corporation of Saskatchewan kindly offered to have a digital version of Figure 5.1 prepared to enhance the quality of this diagram for publication in the book, and I thank her for this courtesy. SSHRC grant number 410-94-1633 made it possible to pay my research assistants and support staff on a regular basis and to cover the other costs associated with travel, interviews, and field work. In the absence of this funding, the project would have remained little more than a dream.

I thank Stephen McBride for encouraging me to consider submitting the manuscript to the series of which this book has become a part, as well as Michael Howlett and David Laycock for entertaining a broad enough vision of comparative political economy to include this work in the series. Gratitude is also owed to the two readers for the University of Toronto Press and the Social Science Federation of Canada, who remain unknown to me. Be that as it is, they contributed a number of valuable suggestions that have made the manuscript a much more user-friendly book. Virgil Duff and Margaret Williams of University of Toronto Press have shepherded the manuscript through to completion from vast distances, no easy feat in itself.

Different parts of the argument have benefited from readings and comments provided on individual chapters and earlier articles by Don Wells, Larry Haiven, Paul Edwards, Robert Lanning, June Corman, and Bob Sterling. These discussions, whether via e-mail, at conferences, or in coffee rooms and homes around the university, were invariably helpful. In addition, I would like to thank Frances Mundy for her able copy-editing and Naomi Frankel for exercising a vigilant eye in proofreading the final copy.

Many friends have helped to sustain me with their good humour through the trying times of a 'more with less' world and its impacts upon university life. They include Jim Waldram, Lesley Biggs, Michael Gertler, Joanne Jaffe, David Hay, Patience Elabor-Idemudia, Alan Anderson, Pamela Downe, and Dianne Relke. In both struggle and fun, these have been good friends indeed, as have Jo-Anne Lee, Adrian Blunt, and John Shields from further afield.

The final stages of manuscript preparation have been done at the Centre for Research on Employment and Work at Griffith University in Brisbane. My hosts at the Centre have been generous and invariably helpful. In particular I would like to thank Ingrid Brunner, secretary in the Department of Industrial Relations, and Jackie Taylor, secretary at the Centre, for their willingness to take time away from their other responsibilities and lend a hand in helping me get the manuscript back to Canada in an acceptable form. Labour historian in the department, Bradley Bowden, has been a superb intellectual host in a new academic setting.

Finally, Ariana, Kendra, and Vija have contributed to this work in ways that they probably could not have imagined. Excusing me for late-night interviews, hours at the computer terminal, final editing, and a myriad of other details, while pulling me back into the important worlds of bike riding, piano and swimming lessons, gymnastics, and homework has helped to keep a healthy balance in a busy world.

MORE WITH LESS:
WORK REORGANIZATION IN THE CANADIAN MINING INDUSTRY

1

Introduction

'A Sign of the Times'

On the afternoon of 24 May 1988 workers at the Cory Potash mine on the outskirts of Saskatoon, Saskatchewan, were devastated to learn that in five weeks, over two-thirds of their number would be permanently laid off. The mine/mill operation in question was owned by the Potash Corporation of Saskatchewan (PCS), the world's largest producer of potassium chloride, one of the three principal ingredients in commercial farm fertilizers. The downsizing came without prior warning, or the usual rumour-mill activity. Indeed, only two weeks earlier the company's CEO had made a video appearance before workers with an upbeat message on improving price and market trends in the industry.[1] Not long before, the mine had actually added to its workforce, hiring several new workers.

Overall, 152 unionized jobs were being permanently terminated. By recent Canadian standards this was a modest affair, yet it was also deeply symptomatic of a troubling trend that merits closer scrutiny. Employment at the mine (hourly union plus salaried staff) would shrink from 315 employees to just over 100, with only 69 unionized employees slotted to stay on.[2] Incredibly, such complex operations as the mill, which is as large as a medium-sized, multi-storeyed factory, would be run with as few as two workers. Moreover, these were 'good jobs,' in league with the highest industrial wages to be found in Canada, and not easily replicated in a 'have-not' region.[3] Thus, at one level, a comparatively small lay-off in an era of employment shakedown and industrial restructuring was once again observed. The tale could be retold many times over with respect to the fishing, logging, pulp and paper, steel, and automotive industries which, despite the 'post-industrialists,' still define basic features of Canadian political economy. Yet, this incident should not be passed over too hastily. What followed and, just as important, what did not, deserves further reflection.

Difficult as it may be to believe, severance packages were not in order because the employees had been given the legally required five-week notice of the impending terminations. Non-salaried employees simply had these weeks to begin making alternative earning and lifestyle arrangements, even though the company was unwilling to grant them 'premature' leaves of absence to start looking for new jobs. Those with less than eighteen years' seniority were thus trapped. A skeletal workforce would remain in place at the mine, but production would be shifted over to the refining of small quantities of a high-grade, special niche product, which would become known as the 'white option.'[4]

To many the prospective changes made little sense. Although the industry had experienced poor prices and the build up of substantial inventories in the mid-1980s, this was not foreseen to be a permanent condition. And although the company had lost money in the previous three years, this was also viewed as a temporary anomaly that would soon be rectified, as indeed it was that very year.[5] Other answers for the bad news had to be sought.

One year prior to the lay-off decision, new production facilities at one of the company's other holdings, PCS Lanigan, had come on stream. These included both a mine extension, complete with the introduction of the world's largest underground continuous mining machine, and the installation of a second 'super mill' on the Lanigan site. As a result, production at this division was to be more than doubled over the following decade.

Coinciding with this augmentation of production was the initiation of trade action by the American Department of Commerce against the Canadian industry. For the Canadian producers, the United States represented no less than 54 per cent of total sales in 1987.[6] To avoid possible new tariffs ranging from 9.4 to 85 per cent of product value, Canadian producers agreed in 1988 to introduce a new pricing formula that would not be construed by the American authorities as dumping. The new higher prices in turn allowed for the reactivation of two mines in New Mexico. For union officials and others there was a direct connection between the American trade action and the company's decision to mothball 70 per cent of the Cory operation. Jobs at Cory would be sacrificed to avoid tariff retaliation. PCS would cut back on production to make way for the smaller, higher-cost American producers, while at the same time the international price of potash would be supported at new, higher levels.

Charges of this nature evoked anger in the community. Miners' wives and their children organized a demonstration protesting the lay-offs at the company's headquarters, before marching on the premier's office. Later, a delegation of three women from the protest met in an acrimonious session with the chair of the company's board of directors and again with the vice-president of the company.[7] Demands that the provincial government use its new enabling legislation, the Potash Resources Act, to effectively allocate quotas to PCS's four divisions and the other five producers in the province were rejected by management on cost-efficiency criteria. In short, corporate rationality dictated that it was more desirable to shut one division, while running the other units closer to capacity, than to allocate employment more equitably across company properties. Although at the time this option was claimed to be the 'lesser evil' in that it would obviate the need for lengthy lay-offs at the other properties, the company was unable to deliver on this commitment.[8]

The responses by the miners' families were not out of keeping with the militant history that the Cory operation had exhibited since first opening as a private corporation in the late 1960s.[9] Additionally, although initially blindsided by the actions of the company, the local union did respond with an unfair practice suit based on the technological change clauses in the provincial Trade Union Act.

The existing act required that employees be given a ninety-day notice of any technological change that would have a *significant* impact upon members of their cohort. Once served with such notice, the employees and their representative might compel the employer to reopen the collective agreement for the purpose of bargaining the terms, conditions, and tenure of any employment changes. With the five-week termination notice, the union contended that the relevant provisions protecting workers from unilateral technological change had been short-circuited. The labour force had been denied the right to bargain over the changes that the company was seeking to introduce.[10]

Resolution of the case would ultimately depend upon the demonstration of technological change, and on this point the union appeared, prima facie, to have a good case. For the purposes of the act, technological change was defined as having any of the following attributes:

i) the introduction ... of equipment or material of a different nature or kind than that previously utilized ...

ii) a change in the manner in which the employer carries on the work ... that is directly related to the introduction of that equipment or material;

iii) the removal by an employer of any part of his work, undertaking or business.[11]

For the miners, the third criterion was the most germane. Downsizing at Cory entailed the removal of the manufacture of one line, 'red product,' and forfeiting its production to other sites that would make up the shortfall.[12] Further, the switch over to a new product line at Cory division, the so-called white option, required the introduction of new technologies, specifically the use of different chemical materials and procedures in the milling process. Although the company had undertaken the planning of these changes to allow for the budgeting of the 'white option' as early as autumn 1987, the union contended that it had not been given proper notification under the terms of the act.[13]

Management viewed things differently. From its perspective one product line was being discontinued, while another was being expanded. However, since 'white' product was already being milled on site, a rededication to this commodity market did not entail the introduction of a new technology – that is, technological change – but rather a *modification* of existing technology.[14] In essence, then, company management argued that it was simply scaling back on one product line in response to the state of market demand. A successful challenge to this decision would be nothing short of undue interference in management's right to manage and, specifically, would call into question managerial abilities to decide what to produce.

On the essential question of removing one part of the business to the company's other property holdings and of the subsequent changes that this might entail, the company's argument is worth considering in detail. Advancing a restrictive interpretation of the pertinent clauses in the Trade Union Act, it proposed that the term 'removal,' 'does not mean that an employer who has a capacity to do work or produce product which exceeds his present demand and lays off some of the employees doing that work or producing that product has "removed" the work of the laid off employees to his remaining employees *even if* such remaining employees *do more work or produce more product* than *they would have, had the employer not made any lay-offs* [my emphasis].'[15] According to the manufacturer it was necessary to distinguish between simply '*moving*' a business operation and '*removing*' it. In the company's view, and that of its lawyers, the latter referred to changes that are consequent on the introduction of a new technology, as when a certain stage of the work procedure has been made redundant by the introduction of new equipment. It did not refer to spatial relocation that may be caused by the state of market demand.

Incredible as the semantics and terminological hair-splitting of this argument may appear, it won the day. The arbitration panel concluded that part of a business was being shut down rather than 'removed,' *even though* workers at other sites would now be responsible for producing more of the product that had been abandoned at Cory. And with that, the redundancy of 152 workers, out of a total unionized workforce of 221, was made legally definitive.

Apart from the outstanding nature of the judgment, which in itself involved quite a liberal interpretation of the text of the act, several other features in this episode stand out. Taken together, and often repeated in similar situations elsewhere, they are what lend such events and their outcomes an aura of inevitability in the era of post-Fordism. First, the events at Cory remained localized at one plant within a growing transnational corporation. Specifically, the case brought by the union against the company considered technological change only at the one division where the immediate impact of job redundancy was being registered. However, considerable upgrading had been completed at another division the year before, and it was expected that this operation, and possibly others, would assume the share of 'red production' that had been forfeited by the Cory operation. The connection between technological change at one operation and downsizing at another was, for better or worse, never directly made by the union in its legal brief.

This may be taken as symptomatic of an overall absence of effective liaison among union locals over the crisis in the industry. Indeed, with the abandonment of the joint potash bargaining council in 1988, the unions seemed to be moving away from cooperation with each other.[16] In short, as downsizing began to occur and insecurity spread throughout the industry's workforce, local unions tended to hunker down in the hope that they would be spared from the job-cutting measures they saw going on around them.[17]

This situation was further abetted by the presence of three different bargaining representatives in the industry and a history of isolation among them. The United Steelworkers of America represented workers at five potash mines, including two divisions of PCS-Cory, the site of the radical restructuring, and Allan. The Communication, Energy, and Paper Workers Union had certification at two mines, including the Lanigan division of PCS, the property that had recently undergone expansion. Workers at another PCS property, Rocanville, were represented by a non-affiliated employees' association. Overall, the historical relationship between these organizations had been 'chilly,' with suspicions of employer favouritism in the first mine certification campaigns, and counter-productive competition in later representation elections.[18]

Such divisions did not help in confrontations with management. Indeed, one cannot help but gain the impression that the company had decided to resolve its labour and production problems at each site in turn, beginning with a ten-month-long strike and union defeat at Lanigan in 1987. This was followed by the downsizing at Cory in 1988, and was capped off with ongoing indefinite lay-offs at the Allan division throughout the 1990s.[19] On each occasion, the local union had been left to fend, largely unsuccessfully, for itself.

A second notable feature of the above events is their *ex post facto* quality. That is, notice of the downsizing at Cory was received in the late spring of 1988; arguments pertaining to the unfair labour practice, which was filed over the lay-offs, were submitted in the autumn of 1988 and a decision was rendered more than a year later, in the winter of 1989. Of course by then the job losses were a fait accompli. How likely was it that any tribunal would reverse the company's initial decision and order the reinstatement of two-thirds of the workforce, along with the resumption of production of an abandoned product line? Even if the company had been found culpable, how likely was it that the status quo ante would be restored? Indeed, in this, and similar cases pertaining to job losses and capital mobility, how would the status quo be restored? In fact, the best that the union and its members could have hoped for was a humane severance package that would ease the displacement. In this bid, they were singularly unsuccessful.[20]

These queries point to the time-consuming and reactive nature of the grievance arbitration process as summed up by other commentators in the notion of 'obey now, grieve later.'[21] In pursuing such a process, the union effectively cedes strategic advantage, especially in cases of shutdown or the decommissioning of facilities. Yet it is instructive to note that the grievance strategy was really the union's only rejoinder to the crisis that it faced, even if, as we have seen, it was inadequate. Why, then, were extralegal repertoires, or dual-track strategies not more evident?

The answer most likely rests in the combination of fear and hope, in that order, that had seized the local. The facility was not being closed down in its entirety; only about two-thirds of the operation was to be wound down. No one was quite sure how the vagaries of seniority and skill would play themselves out. Thus, those who thought that they would be retained did not want to jeopardize their chances by participating in such acts as plant occupations or mass mobilizations. These type of reactions, it was feared, could push a hostile provincial government into closing the complex *tout court*. Simultaneously, others did not want to forsake their chances of possible recall in the future.[22] The same logic spread outward to the other locals. Why bring on the possibility of retaliation in such times as these? Together, these attitudes added up to the demobilization that took place. Under these conditions, the legal gesture was the one remaining option. But by placing all of its resources in this measure, the union relinquished both the advantages and risks that might have ensued from a mixed, multi-track strategy of legal and extralegal reply.

In addition to the issues of the localization of struggle and the defensiveness of response, a final point emerges from this narrative that will provide a thematic anchor for the remainder of the book. In its decision on the technological change grievance at the Cory potash mine, the arbitration panel neatly echoed a central managerial contention. The board held that 'an employer who shuts down part of his work for economic reasons has not implemented a technological change by "removing" the work of the laid-off employees to the remaining employees, *even if such remaining employees do more work or produce more product than they otherwise would have* [my emphasis].'[23] This theme, of more with less, is reflected in the choice of title for this book and is explored more closely below. The theme has various dimensions. As suggested in the preceding narrative, more work with fewer workers is one way in which the economic dynamics of the new order are being played out. More work for comparatively less pay, or an enhanced work-effort bargain, is another. These points are examined in much greater detail in this study.

For the purposes of this introduction it is sufficient to highlight that, for the majority of individuals affected by these events, there has been no proverbial 'silver lining.' Rebounding corporate profitability has not had the expected spinoffs at the base. The taken-for-granted relationships between production levels, commodity prices, profitability, and employment have not been sustained. As in other cases where massive downsizing has coincided with equally massive appreciations of stock worth, PCS corporate stock, which was sold off in the 1989 privatization at $18 per share, traded on the Toronto Stock Exchange at $97 per unit in October 1995. Meanwhile the company entered into a major buying spree, beginning in 1990 with the purchase of Saskterra's 40 per cent minority share ownership of the Allan potash mine.[24] In 1993 it followed up with the acquisition of the Potash Corporation of America and its mine properties in Saskatchewan (Patience Lake) and New Brunswick. More recently still, in 1995, there was a significant move towards vertical integration as PCS became a major player in phosphate production with the purchase of Texasgulf Corporation, described as 'a global leader in the production of ... phosphate chemicals.'[25] This left one commodity group in the fertilizer triad in which the company still had no presence – nitrogen production. This situation was remedied in 1996 with PCS's buyout of Arcadian Corp., a U.S.-based multinational and the largest producer of nitrogen fertilizer in North America.[26] With the addition of Arcadian, PCS became not only the largest potash producer in the world, but also the largest integrated fertilizer producer. As if to reinforce the point, the same week as the Arcadian deal was announced, the business press was also abuzz with news that the company had made an offer for controlling interest of Kali und Salz AG, the monopoly that manages eight potash mines constituting the whole of the German industry.[27] Should this deal be ratified, PCS would have more than a foothold into the rich EEC agrarian market. Industry analysts estimate that the company would then control in excess of 50 per cent of global potash trade. [28]

The challenges raised by developments similar to those just described are often encapsulated in such notions as a transition to a post-Fordist, or alternatively, a post-industrial economy. As the debates surrounding these themes constitute the context in which this study was carried out, it is opportune to consider them in a more explicitly theoretical space, prior to entering the domain of field research and survey analysis. This text's purpose is to explore the intriguing paradoxes presented by the contradictory relationships that constitute post-Fordist and post-industrial influences, respectively, summarized here around the theme of 'more with less.'

The Issues of Post-Fordism and Post-Industrialism

One thing that social analysts of different persuasions can agree on is that at some point in the mid-1970s the world as we had known it began to shift in qualitatively significant ways. This continuing change has entailed cultural displacements, geo-political ruptures, and not least, political-economic restructuring.[29] It is with the latter that this book is principally concerned. When the notion of restructuring is invoked, what is being referenced, at least in the first instance, are changing forms of societal organization. Analysts have lent various terms to describe these transformations in social relationships, including globalization, post-industrialism, and post-Fordism. Each is important in its own right in depicting critical alterations in the ways in which economic activity is organized, although for reasons that will become more obvious, greater weight is given to theories of post-Fordism in the analysis that follows.

Globalization refers to new forms of capital expansion that stress mobility and de-emphasize the national identity of business interests.[30] These themes have resonance for this investigation. As previewed in the preceding section, PCS has moved far beyond its status as a junior provincial Crown corporation to become a truly global player. This applies to the other companies examined in this study as well, insofar as they are all transnationals that conduct business on a worldwide scale. In this instance, however, their global character cannot be used to distinguish among them as business organizations. Moreover, spatial (re)location may, or may not, entail changes in the organization of production. Theorists of globalization often make connections between the phenomena they are describing and new forms of work organization, such as post-industrial labour processes, but it is precisely the latter that needs to be investigated rather than imputed to other trends.[31] Finally, although it is true that enhanced spatial mobility has become a defining feature of the new political economy, the mining industry remains tied to place in ways that distinguish it along with other staple industries. As Chapter 3 demonstrates, this industry is dealing with the problem of space in unique ways that are not fully captured in theories of globalization. For these reasons, the theme of globalization is not dwelt upon further in this book.

Post-industrial theory adopts two lines of analysis, one of which is highly germane for this project. As a corpus of literature, post-industrial approaches have considered both changes in the social division of labour (*what* is produced) and changes in the technical division of labour (*how* items are produced). The former considerations focus on the displacement of manufacturing employment by services provision, a theme that is often privileged in such accounts.[32] In some instances, *what* is produced is seen as determinative of *how* such commodities are produced. This can pose a number of problems, however. For one, economic activity increasingly defies classification as solely manufacturing pursuits or services provision.[33] What, for example, are we to make of a mining company which markets itself principally as a service organization and whose overall labour force numbers reflect this designation, as in one of the cases that we will introduce? Or to take a better-known example, are fast-food workers manufacturers, or service providers? Moreover, whether the one or the other, or a combination of the two as in the latter case, some 'new' industries may remain quintessentially tied to old forms of industrial organization, while older sectors, such as those in this study, may take on definite post-industrial hues.[34] As this example highlights, it is simply not possible to read developments in workplace organization and changes in the labour process from shifts in the social division of labour. On this methodological point, Braverman was right; start with the labour process, then move on to the occupational division of labour.[35]

This is not to say that post-industrial approaches ignore production or its social organization. As elaborated upon at greater length in Chapter 5, thoughtful and provocative claims concerning the future of work have emanated from post-industrial theory. Of particular importance here are the trends associated with the automation of work into continuous-flow production processes. Such advancements are claimed by post-industrialists to demand a rejection of previous forms of workplace organization such as Taylorism, with its attendant tendencies to fragment, simplify, and deskill labour. In its place, post-industrial labour processes are said to require analytically skilled workers who thrive in an environment of uncertainty and responsibility.[36] In this fashion the new technologies of the post-industrial era call forth a new, highly skilled workforce, as well as a rehumanization of work environments. These are weighty claims that merit and receive further attention in the body of this study.

As opposed to other topical inquiries, the problem of post-Fordism is first and foremost taken up with the issue of departures in the forms of organization that order social life. This includes the organization of the labour process within production; the organization of what is later referred to as the political apparatuses of production, that is, the norms, rules, and laws that regulate the employment relation more generally; and the spatial organization of production. Thus different regimes of production may well include changing patterns of spatial deployment, as also cited in theories of globalization, as well as new technologies and leading economic sectors as are highlighted in post-industrial approaches. Post-Fordist theories, however, bring these elements back into relation with the labour processes and industrial relations regimes that characterize a given political-economic epoch. As such, a comparison of Fordist and post-Fordist forms of organization provides a valuable framework within which to place developments such as those that occurred at PCS-Cory in a broader context.

The term post-Fordism has as its referent the concept of Fordism, and it is out of the unraveling of the latter that glimpses and debates about the future have come into view. The notion of Fordism has been taken to mean different things – the mechanized assembly line, mass production, the five-dollar day, and the use of semi-skilled labour. Although all of these features are important components, it is more useful to follow the lead of Gramsci and others in denoting Fordism as a specific form of societal organization.[37] Fordism did emerge in the post-1945 industrial societies, however, there was nothing evolutionary about it. Rather the institutions that gave it substance were put in practice to varying degrees and with differing levels of emphasis in different places, depending upon the play of politics.[38]

Nevertheless, Fordism did have a number of key institutional supports that could be manifested in diverse ways. Within the workplace some form of union recognition and collective bargaining was required in order to enforce the productivity and wage accords that permeated the economy. Such labour–management relations could be highly formalized, as under the Wagner-style labour laws of the United States and Canada, or they could remain uncodified but nonetheless well understood, as in many instances of European industrial relations and the de facto cases of productivity bargaining.[39] The existence of organized labour–management relationships at the workplace further entailed an accentuation of bureaucratic tendencies that accompanied the rule of law in employment relationships. Although task subdivision and the principal job concept were first deployed by managers as a means of organizing large anonymous workforces in a 'scientific' fashion, trade unions were later to seize upon the same strategies in the form of job control unionism in efforts to assert their own priorities in legal contracts with employers.[40] In other words, employment relations were thoroughly bureaucratized, although not necessarily harmonized, under Fordist protocols. The quid pro quo for all of this came in the form of the residual rights that were accorded to management. These included the unilateral right to organize the labour process in ways that management desired, or the so-called right to manage, which became a recognized feature of Canadian and American collective agreements.[41] In turn, this right helped to vouchsafe the annual productivity improvements that were part of Fordism's 'virtuous circle.'[42] More specifically, management was ceded the wherewithal to further subdivide and deskill labour, thereby bringing Taylor into the heart of the labour process.[43] Sufficiently high levels of union density, along with transfer mechanisms such as pattern bargaining practices or, in some countries, neocorporatist arrangements of a highly organized variety allowed for a greater generalization of Fordist wage norms.[44]

A second defining moment of the Fordist paradigm was the birth of national welfare states and the extension of the social wages that accompanied them. The presence of such polities and their constituent programs provided the labour market conditions underpinning the high wage economy that was part of the Fordist dynamic.[45] Thus commitments to high levels of employment through active fiscal management made up one pillar of the existent political economy.[46] Various guaranteed income transfers to households, in lieu of wages, formed the other. Along with the diffusion of collective bargaining throughout key sectors of the economy, the Fordist welfare state and Fordist industrial relations provided the social infrastructure for the system of *mass production, low unit production costs*, and *high wages* that ultimately define this form of capitalist production.[47]

Different national societies approximated this paradigm to varying degrees and through distinct modalities. Thus the undertaking on behalf of full employment varied across place. Welfare programs differed with respect to their active and passive qualities, as well as in accordance with their centrality or residual relation to the market sector.[48] Union densities ranged across place and time, from less than one-third of non-agricultural workforces (the United States, France), to between 35 and 50 per cent (Canada, the United Kingdom, Australia, the former Federal German Republic), to over 80 per cent (Sweden). Meanwhile, depending upon past historical struggles, the protocols for collective bargaining could be totally disparate among different Fordist economies, moving from the extreme decentralization of the North American scene (most frequently plant-by-plant, or employer-by-employer negotiations, only occasionally by industry), to the organized negotiations among peak bodies that have been a feature of several European political economies.

A final attribute, which will not figure largely in the following discussion, but which nonetheless has already been alluded to and should be noted here, is the national quality of Fordism. Thus, it is not only possible to refer to specific national variants of Fordism. With its props of labour market and welfare regulation, and ordered industrial relations, Fordism presupposed a strongly interventionist *nation* state.[49] On the other hand, international forms of regulation were less adequately developed and a truly global Fordism was never a reality.[50] This situation could persist as long as one national state was strong enough to enforce a regime of regulation at the international level, to which other states and economic actors found it to their advantage to abide.[51] In practice, this was the Bretton Woods charter with its system of bounded, mutually supportable exchange rates. But the international dimension was also to be the weak link in this system, and once the world currency could no longer hold its value, or credibly play the conflicting roles that had been assigned to it, that system began to fray.[52] Hence, it was no accident that unregulated market forces first began to strongly reassert themselves at the international level, manifested in intensified international competition, balance-of-trade disequilibrium, balance-of-payments problems, and currency crises. These events constituted the precursor to new political-economic-restructuring projects in one national political economy after another and to which the appellations of Thatcherism, Reaganomics, and so on, have since been attached.[53]

Thus, although the contradictions of Fordism first *appeared* in international economic relations, it is in the workplace and associated national labour markets, as the opening of this chapter suggests, where people have experienced the most profound socio-economic restructuring. This has also been the main site of contestation. That there has been major erosion of the Fordist infrastructure, there can be little doubt. Political attachments to high levels of employment have been abandoned, more expeditiously than most thought possible.[54] Coupled with this, there has been a serious downgrading in the status of the welfare state and its support for the unemployed and wageless.[55] Such actions, however, much as they might contribute to the resolution of public debts, are not merely ends in themselves.[56] Rather, the deregulation of national labour markets is part of a larger project entailed in the reordering of production.

The crisis of Fordism has been diagnosed as an interruption in the 'virtuous circle' of mass production – falling unit costs and rising incomes. These class/production relations had reached their limit owing to a number of different factors. Sustained high employment was thought by some to have occasioned greater amounts of indiscipline in the workplace, although, as suggested by the Japanese example, this is at best a partial explanation. For others, the organization of work around the canons of scientific management – thorough-going task subdivision and strict managerial control of production – had passed over into the zone of diminishing returns.[57] That Fordism as an organizational paradigm had run out of steam was signaled by the stagnant productivity levels and bouts of inflation that had become commonplace among the large industrial economies.

To summarize and simplify the argument, Fordism can be defined as a unique combination of *Taylorism* at the workplace, *Wagnerism* in the sphere of industrial relations, and *Keynesianism* in the labour market. The first element, Taylorism, or scientific management, as Braverman has so eloquently analysed, is premised upon the separation of the two vital aspects of work activity: planning and execution.[58] The former is turned over to the domain of management, most often engineers, following a process of expropriation that is made feasible through the methodology of scientific management.[59] Accordingly, thoughts, initiative, humour, playfulness, intelligence, and other dimensions of humanity are best left at the entrance to the employment relationship. The results, predictably, are the production of deskilled and demeaned human subjects who have become accustomed, if somewhat reluctantly, to following instructions. Of course these claims have been thoroughly contested in what has become known as the labour process debate.[60] These issues will be discussed more fully in a later chapter. For the present purpose of theoretically explicating Fordism and the transition to post-Fordism, however, elements of Taylorism will be assumed to have been important to what managers do.

The second element in our theoretical equation, Wagnerism, refers to the national labour codes that govern the conduct of industrial relations. In this case, the reference is to Robert Wagner, author of the 1935 National Labor Relations Act in the United States (NLRA). This measure established a paradigm consisting of procedures for the legal recognition and certification of independent trade unions as well as protocols for bona fide collective bargaining. Nine years later, Wagner's model was largely emulated in Canada and has since exerted an influence on industrial relations reform in postwar Japan, as well as in failed British initiatives.[61] Again though, the use of the term Wagnerism in this text is more generic, referring to the political apparatuses of production that consist of established, ordered, and respected rules (laws, norms, expectations) existing between employers, unions and the state for the handling of industrial relations.

Finally, and better known, Keynesianism refers to the economic, social, and labour market policies that were directed towards the maintenance of income and employment in the world economy.[62] Above all, they countenanced a vital countercyclical role for national governments, along with the creation of supportive international lending organizations.

If Taylorist labour processes, pluralistic Wagner-style industrial relations, and Keynesian welfare measures total up to Fordism, then post-Fordism is, first, a renunciation of this interrelated social infrastructure. At the worksite, post-Fordism is represented by, above all, the search for a new competitive flexibility.[63] This embodies a dismantling of the old, narrow division of labour and a recombination of individual task assignments in expanded, reconfigured jobs. In place of isolated individual workers who are expected to leave their mental capabilities at the employee's entrance to the worksite, full participation in work teams becomes a sine qua non to successful competition on global markets. Thus, if Fordism implied the deskilling and degradation of labour in the work process, post-Fordism, at least as understood by some, stands for a much valued reskilling of work.[64] Significantly, this trend involves not only the new emerging high technology sectors, but also such traditional undertakings as steel, textiles, and automobile production, as well as machine tool manufacturing.[65] In this scenario, then, the survival of individual economic units, as well as the jobs they create, is said to depend upon new *commitments* to the 'continuous improvement' of organizational, productive, and delivery technique, 'total quality control,' defect-free craftship, and reliable, just-in-time schedules. All of this, in turn, requires the creation of a workforce with post-Fordist sensibilities, one committed to the above goals.

Such alterations in the labour process are unlikely to occur spontaneously. Indeed, the model of flexible post-Fordist accumulation goes against much of the grain of pre-existing relations. This is evident when it is realized that Fordism embodied a complex melange of specific worksite and industrial relations controls, including adherence to the philosophy of scientific management, an abiding faith in the efficacy of technological systems, and a penchant for bureaucratizing the employment relationship.[66] In other words, that aspect of the labour process debate that focuses on the narrow question of what form of control predominates at any particular point in time (scientific management versus responsible autonomy, for example) misses the point that Fordism is a complex ensemble of complementary relations.[67] Its deconstruction entails more than simply substituting one form of managerial control for another and therefore reaches into the wider realm of production politics.

As a result, post-Fordism does not merely entail the simple issue of removing Taylorism and implanting new forms of functional or numerical flexibility in individual firms.[68] For as Taylorism is actually encountered in existing Fordism, it is within structures of job classification, and unique payment and motivational systems. These features, in turn, rely upon specific forms of unionization, collective bargaining, and labour–management relations.

A cardinal principle of Fordist trade unionism has been the objective of taking wages out of competition. This objective has been historically effected at both micro- and macroeconomic levels. At the level of the firm, job control unionism has been practised. That is, it is the specific job, not the person filling it, that determines the rate of pay.[69] The latter issue is determined by the application of rules of employment seniority. Thus, each job has a specific rate of pay attached to it, as befits the bundle of qualifications, training, and skills that define it. Lines of progression are marked out, creating what amounts to internal, or firm-specific, labour markets. One of the local union's principal tasks is to be found in the policing of these rules; participating in job evaluations, upholding seniority rights, and taking on the many individual grievances to which the rules may give rise.

At the macro level, taking wages out of competition entails the negotiation of common wage patterns within national economies. This may be accomplished in a variety of ways, including the centralized peak bargaining and neocorporatist arrangements that have been a feature of several Western European economies, or the practices of patterned, sectorial master agreements that are found in North America. In the latter specific collective agreements are concluded with large employers and then used as benchmarks for other unions in other industries. In this manner, the linkage between productivity growth, real wage adjustments, and mass consumption is maintained. This presupposes both strong national labour organizations (as measured, for example, in levels of trade union density) that can realistically take wages out of competition, and a healthy local union presence, wherein workplace job controls are actively enforced. Undeniably then, these are the main features of a pluralistic and adversarial regime of industrial relations.

It is precisely many of these features, and the philosophy that lies behind them, that have become anathema to critics of the status quo, both liberal and neosocialist alike. To the rigidities of the technical division of labour are added another layer of job rules and measures that allegedly invite abuse. In short, such job control militates against cooperation in the workplace, while driving wedges between effort, results, and rewards. It is precisely these structures and practices, ranging from the organization of work at the local level to the practice and results of collective bargaining economy wide, that post-Fordist practices and politics have sought to dislodge.

Beginning again with the labour process, the recomposition of work tasks entails the nullification of existing job classifications that reflect the Taylorist division of labour. The exercise of knowledge and responsibility on the job requires levels of discretion and cooperation that cannot be attained within existing job structures. Ergo, the goal becomes one of creating broadly defined tasks, to be undertaken by integrated, responsible, problem-solving teams whose practices are informed by an ethic of continuous improvement.[70] This redefinition of work has a number of consequences, starting with the role assumed by the principle of seniority. This standard of Fordist industrial relations retains considerably less currency in the post-Fordist environment, where analytic problem-solving skills and quick adaptability to new conditions ideally take precedence in work-team membership and task assignment. By the same token, an effort is made to directly link remuneration with effort and results, through the introduction of payment-for-knowledge systems and new designs for gainsharing.[71]

The same thrust is also evident at the macro level of industrial relations regimes, where the flexibility of enterprise bargaining is substituted for industry agreements and common patterning. In this instance, each unit of production is treated as a 'profit centre' of value-added activity in its own right, while wage levels must be tied to achievement at this level. Ultimately, all of this assumes an essentially new labour–management accord, fundamentally different from that which oversaw Fordist expansion. Adversarial industrial relations are a luxury that can no longer be afforded. In their place, post-Fordism promotes a new partnership based upon mutual understanding and joint problem solving in the search for value-added production.[72]

It is not my intention to elaborate upon the implications of post-Fordism for the Keynesian welfare state, the third element of the Fordist accord, as this has been done by others. However, many of the same themes appear in this element as in the critique of industrial relations. The most obvious includes elaboration of the so-called rigidities that passive income and labour market policies bring to bear on labour force activity. Problems that are thought to have been created by the Keynesian welfare state, such as induced dependencies, malingering, and dysfunctional attachments to specific economic sectors, are to be righted by a more tightly qualified system of benefits, which again ties receipt of transfers to participation in training schemes and workfare.[73] This requires nothing less than a total revision of our notions of welfare, citizenship rights, and social wages.

This study concentrates on the first two-thirds of the Fordist/post-Fordist equation – the labour processes and industrial relations/production politics of the respective regimes. Interestingly, it has been much easier to place a positive gloss on at least some of the changes occurring in these realms as compared to the punitive picture of welfare 'reform.' Few sociologists would be prepared to argue on behalf of the benefits of Taylorism and scientific management! Even after the patent excesses of this system had been confronted by recognized trade union power, there is no doubt that it left much to be desired, as social critics have correctly pointed out. Thus to many, including some on the political left, certain aspects of post-Fordism have held promise, just as certain ascribed features of post-industrialism have evoked positive receptions. Some proponents of a regulation approach have consequently distinguished between good and bad types of flexibility.[74] Accordingly, an intelligent response to the crisis of Fordism will aim at discouraging the emergence of the latter, including the proliferation of insecure, contingent work, while actively encouraging initiatives aimed at training, reskilling, and multiskilling.[75] In very similar terms, Piore and Sabel have looked forward to the promotion of flexible specialization, which in their view brings a measure of craft skill back into work performance along with the possibility of greater worker control over the workplace.[76] These benefits hold for the flexible specialization theorists, regardless of whether they take effect in a unionized or non-unionized setting.[77] Analogously, the post-industrialists see a brighter future ahead as automated production techniques render Taylorism and all of its accoutrements increasingly dysfunctional.

Critics of the post-Fordism thesis are of two types. First, there are those who reject the notion itself, and second, there are those who, although accepting the reality of post-Fordism, dispute the effects that have been imputed to it. A number of studies, principally of British origin, fall within the first category.[78] Generally, these works attribute changes in existent industrial relations to the deep recession of the early 1980s, rather than to transformations in the capitalist order. With the benefits of hindsight, such arguments are unconvincing. The time span covered by these studies is a matter of years, rather than decades, which alone would tend to privilege the cyclical over the structural. They are limited to one country rather than to a highly integrated system of political economies. And finally there is a tendency, best displayed in Pollert's work, to conflate a critique of the claims of the flexibility theorists with the denial of any new developments whatsoever. Changes in the nature of Fordism are undeniable; it is the meaning and implications of those changes that are up for debate.

It is this latter note, the meaning of post-Fordism, where argument has been sharpest and to which this study hopes to contribute. Once again, beginning with the labour process and working up to industrial relations' protocols, the optimistic conclusions that have been arrived at by flexibility theorists, as well as by advocates of lean production, have been contested by other findings that equate post-Fordism with work intensification rather than with skill enhancement.[79] More pointedly, some have identified developments in the sphere of production with a new form of 're-Taylorization,' in which the principles of job study are now applied by workers to themselves, in the name of continuous improvement and other managerial mantras.[80] This is really the crux of the debate, for others have been equally adamant in rejecting equivalencies between Fordism and new forms of work organization that are held to offer better alternatives for all participants in the employment relationship.[81] For the critics, so-called systems of 'management by stress' combine both the tenets of Taylorism with the jointness of union/management cooperation in new regimes of intensified exploitation. For the enthusiasts, on the other hand, post-Fordist organizations are superior in all ways, combining better jobs with more efficient productive organization.

If the potential effects of post-Fordism have been disturbing within the labour process, it has also been passionately argued that they have had devastating effects on the post–Second World War system of industrial relations. Although Panitch and Swartz did not adequately distinguish between the final manifestations of the industrial relations embodied by Keynesianism, and later post-Fordist realities, the great merit of the various iterations of their work is to alert readers to the shifting balances of power in employment relations.[82] Thus, as both Panitch and Swartz, as well as McBride, demonstrate, the point is not so much about the frontal dismantling of Wagnerism; the skeletal framework still remains.[83] Rather, what is at issue is the outflanking of the Wagner paradigm, first by governments, as Panitch and Swartz emphasize, but also through new business practices that make use of the contracting out of operations to non-union suppliers, the greater use of contingent labour, and the planned downsizing of operations along with the consequent whipsawing of employment conditions between individual worksites. These practices, when placed in the context of smaller more diffuse workplaces, present new challenges that simply were not foreseen by the policy framers of Wagnerism.[84] As a result, the legal framework of Wagnerism, which sanctions union organization, certification, and collective bargaining shop by shop, presents labour with few resources to deal with the new industrial relations climate.[85] This in fact may well be one reason why the formal structures of Wagnerism still persist in both Canada and the United States. It may be that they are simply less relevant than once was the case.

The results have been stagnant union densities and real wages as in Canada, or declining real wages and union densities as in the American case, depending upon such factors as the respective union strategies that have been adopted, pre-existing technicalities of the law, and the availability of on-shore non-union settings.[86] In short, critics suggest that Wagnerism has not sufficed to save the essential elements of Fordism, including productivity-based wage structures, connective bargaining patterns, and permanently ascending living standards.

Workplace restructuring also implies a different role for unions at the local level than is found under Fordist production conditions. The disbanding of job classifications, and associated rules of seniority, along with the introduction of new payment plans (gainsharing, payment-for-knowledge), is viewed by some as promoting a much diminished union presence in the workplace.[87] Employment relations are further individuated through the new payment and promotion systems, while unions are less able to defend members once old job rules have been abandoned.[88] Accordingly, and apart from functioning as conduits of cooperation with management, there really is not that much left for unions to do in such circumstances. This, however, need not imply the advent of more democratic, egalitarian work relationships as has been suggested in other analyses.[89] If anything, critics suggest, this aspect of post-Fordism is more conducive to new forms of company culture and the accompanying practices of employer unilateralism than to meaningful workplace empowerment.

This, then, is the debate. What outcomes are more likely to emerge from post-Fordist work organizations or post-industrial technical frameworks: reskilled jobs in more participatory work environments, where unionization may, at best, become a secondary or even redundant concern for workers; or leaner, more stress-laden production systems, embodied within a new authoritarian corporate culture that, at most, is prepared to come to terms with new forms of enterprise unionism?

The Study: Genealogy, Design, Accident, and Politics

It is not realistic to expect a full-scale resolution to all of the issues that have been raised in the foregoing debates in one, albeit comparative, empirical study. Ultimately, attention can be drawn to some features of important trends, and in the process some poor arguments may be rejected, while some good analyses may be qualified and improved. Further comparative study (cross-industry and cross-national) will definitely be required to confirm or reject the competing hypotheses that are associated with post-Fordist and post-industrial phenomena. Nevertheless a comparative sectorial study provides a promising starting point.

Why select a staple industry, and mining in particular, for study? Several considerations have entered into this choice. First, as numerous others have demonstrated, the salient feature of Canadian industry and its Fordist variant has been its staple, export-based foundation.[90] More generally, staples production has been and continues to be a defining feature of Canadian political economy.[91] In the case of this region, the mining industry has become an increasingly important part of a provincial economy, as well as a link to broader, transnational agro-industrial complexes. Although mining itself is an old, indeed an 'original industry,' the worksites that are analysed in this book all trace their origins from the late 1960s up to the early 1980s. In other words, that these are multinational corporations that bridge the transition between the Fordist and post-Fordist eras while employing both mechanized and highly automated technologies were also important considerations. Of course, not all corporations have handled the challenges that this implies in the same manner. That is precisely what makes a comparative study of different employers in the same industry highly instructive.

When I first began to conceptualize a study of the mining industry, my interests were more restricted. It appeared that one of the more novel innovations that had been introduced into the industry is what I will term 'radically spatialized' industrial relations. This involved a break from past practices whereby workforces were resettled in the vicinity of an operation and would often remain there for several generations of production.[92] Currently some forms of mining operation have planned lifespans of no more than twelve years, thereby providing far less employment continuity and job security than was once the case. In these circumstances, companies have drastically altered the employment relationship by instituting long-distance commuting patterns for their workforces. In other words, both the turnover time of capital and the temporal and spatial dimensions of the employment relationship have been radically altered. Where once the industry had provided long-term employment in spatially fixed communities, now short-term work in *non-existent* communities was emerging in certain sectors of the industry. These alterations seemed to capture in microcosm many of the trends that were beseiging the larger economy. Given a previous history of sociological investigation, I wanted to examine an apparent consequential development.[93]

The prototype for such spatialized employment relations was the uranium industry, where radically spatialized work relations are now the norm. The Canadian industry is located almost entirely in northern Saskatchewan, where it is regulated by provincial surface lease agreements that encompass, along with other stipulations, an affirmative action program for northern residents. As a result, the workforce in the industry is composed of First Nations' members who reside in both the north and the south of the province, southern whites, and a comparatively high proportion of women workers. Yet this was clearly not another example of the migratory labour system that, for example, had characterized the mining industry in southern Africa.[94] Here, the defining moment was the rapid turnover time of invested capital, which did not allow for the establishment of even semi-permanent settlement anywhere near the worksite. A new form of employment regulation was imposed, characterized by weekly work stints at the job site, followed by weekly periods of life away from work in diffuse home communities.

Space and turnover time have not been prominent dimensions in political-economic analysis. One noted exception to this, however, has been David Harvey's work, which has brought these relationships back into play.[95] For Harvey, it is precisely changes in the interrelated spatial and temporal modalities of capitalism that distinguishes the post-Fordist phase of 'flexible accumulation' from the Fordist era of mass production. A study of the labour relations in the new uranium industry, with its just-in-time workforces, appeared to offer an opportunity to explore, test, and push these propositions further.

The largest uranium producer in the world is Cameco Corporation, which had a previous life as Eldorado Nuclear, a Crown corporation.[96] My original intention was to conduct a general workforce survey of unionized Cameco employees in the company's largest mine/mill operation at Key Lake, Saskatchewan. Specifically, it would hone in on the defining feature of work at the company, the potentially problematical relationship of managing a permanent and highly paid migrant workforce. Following previous examples, corporate management was approached for permission to conduct an industrial relations study.[97] Initially, as the principal researcher, I was turned down and the project was rejected.

Blocked from this side, I decided to approach the union that had the Key Lake certification. There had recently been a change of staff representatives in the union's (United Steelworkers) district office. The new acting representative was young, intellectually curious, and quite concerned about the implications that changing human resource policies were having on all of the locals that he was responsible for. As Chapter 4 describes, his home local at Cominco Fertilizers (Agrium Ltd.) was being approached about cooperation in a continuous improvement program at the worksite.[98] At the time, the steelworkers were responsible for five potash operations in the area, as well as the Key Lake uranium site. Each of the four companies involved was experimenting to varying degrees with, or had moved to introduce, significant modifications to employment regimes that had been in place since the early 1970s. The union was obviously interested in the implications of these developments, both on behalf of its membership, as well as at the institutional level of maintaining an ongoing presence in the workplace. Although the steelworkers had drafted a general set of principles on union participation in work reorganization, individual locals still seemed to be hesitant about how to respond to managerial initiatives when confronted with them. To participate, and if so, how far; or to resist, and again in an active or passive mode?[99] These were and continue to be the issues of the day for union locals.

This level of union interest then, was the beginning of a much broader project than originally had been conceived. Five local unions represented by the Steelworkers agreed to participate in a study that would focus on comparative work regimes at four multinational mining companies.[100] Four of the sites were potash mine/mill complexes, while the fifth remained the original uranium mine/mill operation. Each site was, in its own manner, the object of new managerial initiatives. The union was keenly interested in drawing forth any data that would shed light on the human impacts associated with organizational and technical change. This pressing concern dovetailed nicely with anchoring my own focus on post-Fordism and post-industrialism in an important empirical reality. Out of these joint involvements the project was born.

The first task was to gain a basic qualitative familiarity with each site, including the labour processes and, in particular, with the industrial relations that stood out as defining features at each complex. Although labour processes at each of the four potash mines were essentially the same, management strategies for obtaining and maximizing the work-effort bargain varied considerably.[101] This seemed to be the point to hone in on.

As part of the initial 'familiarization phase,' a number of site visits were paid to each company, except Cameco, the uranium producer. On these occasions at least part of a shift was spent underground and in the mill to gain a basic appreciation of the different work tasks and technical processes that were being carried on. Out of these visits and the lengthy discussions that occurred prior to, during, and after them with key informants from the local union executive and interested members, a draft survey was produced. This document went through numerous versions with input from local activists and two different staff representatives before being pretested with volunteers from one potash mine and the uranium site. Throughout the process, members were asked what questions they were particularly interested in and were invited to place them on the survey. At the same time, they were asked to comment on the relevance and sensitivity of my queries. The pretesting was also an exercise in additional information gathering. Some questions were intentionally left open-ended to see what type of responses would be elicited. On the basis of later content analysis, these questions could be further refined, elaborated upon, or 'closed off.' Following the pretests and further consultations, more versions were produced, until a final document emerged on iteration number fifteen!

Before starting the workforce interviews at any given site, the local union was asked two simple questions: 'what do you think is the most important managerial initiative or program that affects labour relations at the mine?' and, 'do we have any documentation on it?' These questions were intended to augment the preparatory qualitative work that I had conducted on variations between the sites. This turned out to be an especially apposite line of inquiry. Since most managerial initiatives, including scientific management, are premised upon worker awareness of and cooperation with the basic features of the program, documentation is surprisingly widespread and available. In all cases, whether it encompassed company program and implementation manuals, training modules for managers, or minutes from joint union/management steering committee meetings, documentation on corporate managerial programs was an extremely rich source of information. To complement this line of inquiry, I also requested and was given access to the local union minute books and any other internal documents that the locals had compiled. In some cases, these documents provided clues as to how workers through their unions had responded to changes in the workplace.

Of course, documentation of human resource management initiatives is one thing while their reception in the workplace is quite another. With that distinction we pass over from the realm of text, to the sphere of ongoing collective interaction. Methodologically, this was signified by the beginning of the labour force surveys that covered a two year period, from July 1993 to July 1995. Respondents were selected from local union membership lists, but the sampling frame at each site was stratified to reflect the departmental/occupational composition of each operation. This was to avoid an inadvertent over-representation of a particular occupational group at any one site, which could skew the results. Thus, the workforce at each mine was occupationally located (for example, underground production department, underground maintenance, et cetera), and random samples were selected from each department in proportions that reflected the department's contribution to the total on-site labour force. Covering letters, informing the selected sample of the study and requesting their participation in the survey, described this as an independent research project with which the union was cooperating. The actual interviews were conducted at the respondents' home addresses, except occasionally when they were more conveniently done in an office at the local union centre, or in my office at the university.

As one might expect, during the interviews, new and puzzling information was often turned up. Thus, the key informant interviews became an iterative process involving current and former members of local union executives, the Steelworker's staff representative, employees who had received special training in company programs, or those who had taken specialized union training in areas such as job evaluation analysis.[102] These were entirely open-ended interviews that served three functions: to clarify and supplement information that was being collected in the labour force surveys; to keep the author abreast of current developments at each research site; and to keep the project visible and vibrant on what were, undoubtedly, crowded union agendas.

As part of this dimension of the study, I did take part in several 'significant events' that were associated with particular locals and that will be discussed more fully in the following chapters. This included discussions on Employee Involvement programs in the local unions, meetings to discuss job actions around new management initiatives, contract ratification votes, and the various bargaining committee and shop stewards' meetings that were an integral part of these discussions. My inclusion in such events signified the growth of a relationship based on trust and mutual respect that I hope is evident in this book.

A final aspect of the study were open-ended interviews conducted with management at each company. These provided an additional reality check, in which mine managers were queried about human resource initiatives and their results, in much the same fashion that the local unions had orginally been approached for documentation. In certain cases, notably with Agrium Inc. and Central Canada Potash Ltd., contacts became more extensive. Agrium permitted me to sit in on company/work crew meetings in which a new Peer Review system was being proposed as part of its Continuous Improvement initiative. Both Central Canada and Agrium provided me with access to internal employee 'climate surveys' that they had previously conducted, as well as with associated documents.

When the workforce interviews were completed at each locale, a first pass was made through the data at that site. Short descriptive 'highlight' reports were authored for each local and every person who had taken part in the survey received a copy, along with an invitation to forward any amendments. Just as we were reaching this stage of the project at Cameco, the last firm in our survey, management, which had previously vetoed either site visits or other forms of participation in the study, came through with permission to visit their northern Key Lake site. This allowed me to fly in on the company's chartered jet and spend two days touring the camp and mine facility, as well as to conduct interviews with managerial personnel on site. Additional interviews with Cameco management were conducted in the south, following this excursion.

The study unfolded, then, partially by design and partially through circumstance. Although I had first given thought to doing the labour force surveys on workplace premises, and hence on 'company time,' this proved impractical, and fortunately so. Clearance to go ahead with on-site interviews would have required corporate approval in four different cases. Attaining this level of cooperation would have proven very difficult and could have complicated the high trust relationship with the union.[103] Although *in situ* interviewing would have been more economical, especially in the case of the northern uranium mine, the final product would have been quite different. Interviewing on company time would have placed greater constraints on the length and inclusiveness of the survey instrument, consciously or otherwise. As well, witnessing the problems that were experienced by two of the outfits in conducting their own, on-site climate surveys convinced me that there is no substitute for totally voluntary participation. In other words, *in situ* interviewing had the potential for posing problems around issues of privacy, confidentiality, and coercion, as these are experienced by workers. Taking the project off of corporate premises and into workers' homes helped to mitigate such concerns. In the end, interviewees participated because they wanted to, not because they were expected or told to. Participation may have been based on hopes for improving the situation at their workplaces, curiosity about the project, or other motives. What was important was that I and the other interviewers remained the guests of the people who were being interviewed – in their space.

In one sense I did not take leave of any of the research sites that have been covered in this study. Rather as investigations at one mine continued, research at another was begun. Comparison was thus firmly interwoven into the very fabric of the study. The succeeding chapters follow for the most part the same logic as the survey itself and the methodology that informed it, with the field research used as the critical context.

Chapter 2 examines the product and labour markets in potash and uranium, while providing an overview of the industry as a whole. A detailed, qualitative analysis of each of the four firms is presented in Chapters 3 and 4, where the main variants of post-Fordism are identified. Although each of the workplaces examined in this book reveal aspects of post-Fordist work relations, some do so to considerably greater degrees than others. The first two companies examined, which have what I have termed radically marketized and radically spatialized industrial relations, are at the low end of the post-Fordist trajectory. The third firm has entered upon a post-Fordist path more deliberately, but still in a decentralized and somewhat ad hoc fashion. Finally, compared with the others, management in the last case has demonstrated a more thoroughgoing knowledge of and commitment to a post-Fordist alternative.

The qualitative insights that are derived in Chapters 3 and 4 are then put to work in an inspection of the labour process in Chapter 5, including the all-important debate on work reorganization and the skill implications that inhere in it. The claims that have been put forward in defence of post-Fordist and post-industrial arguments, as well as the critiques of both are further scrutinized in this chapter. Chapter 6 expands the picture somewhat to examine the impact of employment regimes on industrial relations in the workplace. Do innovations in the organization of work present even limited opportunities for worker empowerment, or are they more likely to lead to just the opposite outcomes? The traditional notion of industrial relations is expanded to encompass all of the political apparatuses of production, as well as the diverse production politics associated with them. The analysis is therefore widened to include topics such as occupational health and safety, job stress, and workplace harassment as integral features of production politics, along with the more standard fare of discipline and grievance handling, and levels of employee input into decision making. Chapter 7 concludes with a brief examination of the impacts of contemporary employment relationships in the industry on civil society as well as a discussion of the overall findings. Before moving on to the wider theoretical issues and debates, however, the outline of the uranium and potash industries that follows provides a background for the remainder of the study.

2

Market Preliminaries: Product and Labour Markets in the Potash and Uranium Industries

Over the last twenty years, the unleashing of market forces in one political jurisdiction after another has contributed in some circles to the credence of a form of 'market determinism' in social analysis. According to such approaches, social relationships, indeed society itself, is reducible to the laws of the market place. In one infamous quip, society itself no longer exists. In a stunningly unconscious, yet ultimately erroneous inversion of Karl Polanyi, only the immutable laws of the market have anything other than a fictive existence in today's world.[1]

The inadequacy of such a position is particularly obvious in the study of large-scale organizations such as corporations, where much social activity is internalized within the organization itself. The organization of work by managers and the struggles around the work-effort bargain are a case in point. These activities are undertaken within labour processes, prior to finished commodities hitting the market, or in the case of labour power, after it has been purchased and removed from the market. For this reason one focus of this study is on 'cultures of employment'; specific strategies that are directed to managing workforces and influencing the work-effort bargain. Even in those cases where efforts are made to import market forces into the social relationships of the labour process, as in the new paradigms that scope so-called internal customers and that intensify outsourcing, such conventions remain social constructs, that is, deliberately designed ways of doing work. In other words, the complexity of labour/management relations certainly exceed the sum of the market dispositions at any given time.

Thus, although the social organization of work cannot be entirely dissociated from the commodities that are being produced, or the markets for which they are being produced, it is certainly not reducible to them. To claim otherwise would be tantamount to privileging the abstract 'laws' of the marketplace and technological development over the real human decisions that are made around what commodities to produce and how best to produce them. It would also entail short-circuiting any consideration of the contentions that such choices often give rise to. Although production is organized *for* the sale of specific goods on well-known commodity markets, it is not uniquely organized *by* such markets in any direct sense. Nor do the specific nature of the commodities under consideration *singularly* determine the labour processes or industrial relations around the productive effort, although they do, nevertheless, exercise some influence, or are susceptible to being used as a resource in their own right, in the management of work. This may be registered both in the unique physical environments where the products are found and appropriated from nature, and in the product markets and terms of trade under which they are sold. Thus, we begin by examining the product markets, or ultimate destinations, that the productive activities are directed towards, and with the labour markets through which specific types of labour power are recruited for employment. This is a necessary preliminary to the following analysis of the cultures of employment and their adherence to specific methods of management.

Potash and uranium are commonly referred to as staple commodities. As those who are at all familiar with the annals of Canadian political economy will recognize, the concept of the staple has become a charged descriptor. That is, over the years, there has been an intense debate on the contribution of staples analysis to social theory. Staples theory has come to represent different things for different analysts.[2] Therefore, it is incumbent to state at the outset exactly what the status of references to staples production are for this study.

Staples analysis has been used in three somewhat different ways: as a master narrative of national economic development or economic history;[3] as a sociology of social class analysis;[4] and as a specification of certain types of product market/price formation behaviours. Although these three dimensions are obviously related to one another, the notion of a staple commodity has relevance for this study only in the third and final sense. This is simply a recognition that certain forms of market exchange and product demand may in turn be used to affect the work relation itself. Specifically, in the case of the industries studied here, product demand is largely international in character, while commodity price is highly variable. Production is chiefly for export, and prices are set through international commodity markets. As a result, certain forms of economic regulation and the absence of others, distinguish these industries from other undertakings. It is therefore necessary to begin with a study of the markets – the product markets for these commodities, and the markets in labour power that they consume. Although important, these aspects are only a beginning. They lead to a consideration of the economic units that produce potash and uranium for the international marketplace, and to the managerial strategies and corporate cultures that organize the workplace. As we shall see, relatively homogenous product and labour markets may still lead to a plethora of employ-ment relations in varying cultures of employment.

Use-Values and Exchange-Values: Markets and Prices

Potassium chloride (KCl), or potash, is one of three basic agents used in the manufacture of commercial fertilizer. Where it does not occur naturally in the soil, it must be introduced through the application of blended fertilizers that contain, along with potash, specified quantities of nitrogen and phosphate as determined by local soil conditions. Potash, then, is a basic component of the agro- industrial complex that dominates world food production. It is used to increase crop yield, enhance plant resistance and durability, and augment food value. There are no known natural or artificial substitutes for it.

Since 1992, and coinciding with the collapse of the former Soviet Union (FSU), Canada has become the world's largest producer of potash, as well as the largest exporter.[5] As indicated in Table 2.1, Canada currently accounts for well over one-third (36 per cent) of global production, followed by the FSU (Republics of Russia and Belarus), Germany, the United States, and Israel. The total KCl production of the latter three states still does not amount to three-quarters of Canadian output. Within Canada, production is largely centred in Saskatchewan, where most recently ten operating mines have accounted for 88 per cent of national production (1994). The remaining 12 per cent is produced by two underground mines in New Brunswick.

Although knowledge of extensive potash reserves in the Elk Point Basin were brought to light in early century geological surveys of the province, potash was only 'rediscovered' during the 1940s search for oil on the prairies. Com-mercial development, however, was quite slow for a number of reasons. The initial inclination on the part of the governing CCF was for full public sector development of potash reserves in a joint venture with the federal state.[6] Fed-eral politicians and planners, however, did not share their enthusiasm, and with-out the assurance of such participation, provincial politicians turned to the private sector for development capital. Following a number of unsuccessful negotiations between 1948 and 1952, the Potash Corporation of America was finally lured into the province with a series of generous concessions on royalty rates that would become the model for further development. The first mine opened for production in 1958, although it quickly experienced problems with flooding. The first continuous production operation came on stream only in 1962, but it was quickly followed by what can only be described as a boom in

TABLE 2.1
World potash production, 1989–94 (000 tonnes K20)[a]

	1989	1990	1991	1992	1993	1994
Brazil	109	98	101	85	170	230
Canada	7,333	7,002	7,405	7,270	6,850	8,150
Chile	20	20	38	35	35	35
China	32	46	60	60	60	60
F.S.U.[b]	10,232	9,126	8,510	6,948	4,667	5,090
France	1,195	1,292	1,129	1,141	890	890
Germany	5,386	4,850	3,902	3,525	2,860	3,280
Israel	1,273	1,311	1,270	1,296	1,342	1,300
Italy	154	68	31	86	–	–
Jordan	792	841	818	808	822	925
Spain	741	686	585	594	661	685
United Kingdom	463	488	494	530	555	580
United States	1,580	1,654	1,692	1,658	1,525	1,425
Total	29,310	27,482	26,035	24,036	20,437	22,650

[a]Potassium nutrient in metric tonnes which equals 1000 kilograms
[b]Former Soviet Union
Source: *Canadian Minerals Yearbook*, 1994.

mining development. By 1970 ten mines were operational in the province: two were Canadian controlled companies, six were headquartered in the United States, and the remaining two were of French/German and South African parentage respectively.

The much touted partial nationalization of the industry was carried out in 1975, right in the midst of the general commodity price boom that coincided with the first oil crisis. Nationalization was propelled by a number of coinciding factors, including escalating primary commodity prices, the unfavourable royalty agreements that had been signed with all of the producers in the preceding decade, and a series of court challenges to new provincial taxes that had been levied by the provincial NDP in an attempt to enhance stagnant royalty revenues.[7] Although the Saskatchewan government was initially accused by U.S. congressional representatives of OPEC-style behaviour, the political and legal entanglements quickly subsided with the generous compensation awards that were made to the producers. Within three years of the first purchase of the Duval mine from American-owned Pennzoil in 1975, the newly created Potash Corporation of Saskatchewan (PCS) operated four divisions with a 40 per cent share of provincial production.

TABLE 2.2
Canadian exports of fertilizer potash, destination by percentage share, 1993–4
(tonnes)

Potassium chloride	1993	1994
United States	63.9	56.3
People's Republic of China	6.7	13.3
Brazil	3.3	6.1
Japan	4.9	4.0
Malaysia	3.7	3.6
South Korea	3.7	2.9
Australia	2.9	1.9
India	0.5	1.7
France	0.4	1.6
Taiwan	1.2	1.3
Belgium	0.3	1.2
New Zealand	1.6	1.0
Chile	0.7	0.6
Colombia	0.8	0.7
Indonesia	1.0	0.6
Philippines	0.4	0.4
Thailand	0.5	0.3
Cuba	0.1	0.4
Denmark	0.3	0.4
Guatemala	0.3	0.2
Dominican Republic	–	0.2
South Africa	0.2	0.2
Netherlands	0.2	0.2
Jamaica	0.2	0.2
Ireland	–	0.1
Bangladesh	–	0.1
Argentina	–	0.1
Italy	0.3	–
Nigeria	0.2	–
Singapore	0.1	–
Costa Rica	0.2	–
Venezuela	0.3	–
Norway	0.2	–
Mexico	0.5	–

Source: Calculated from *Canadian Minerals Yearbook*, 1994.

Although nationalization could capture a larger share of the rents accruing from natural resource production for the public bourse, it could not alter the product markets for the commodity in question, markets that define potash as a quintessential staple good. Thus, the industry is almost completely reliant upon external markets. In 1994 for instance, domestic sales accounted for only 4.6 per cent of tonnage sales. Consistently, the largest market is found due south, in the United States, which, in the most recent year for which statistics are available, accounted for 57 per cent of the export market as measured in tonnage sales (Table 2.2). Potash is particularly important to the giant American corn and soybean industries. Almost all of this demand is met by Saskatchewan production sites.

After the United States, the largest export markets for potash are to be found in Asia (29 per cent of sales), followed by Latin America (8 per cent), Oceania (3 per cent), and Western Europe (3 per cent). Although the list of importing countries is long, eight nations alone (the United States, China, Brazil, Japan, Malaysia, South Korea, Australia, and India) account for the bulk of all exports. In the last year for which data are available, 1994, exports of potash were worth $1.6 billion, a significant sum by any account.

If one examines the past export record, great variability in final demand is evident. America stays on top as the most important customer for KCl, but its demand can fluctuate considerably from year to year, depending upon factors such as the weather or state supply management policies pertaining to such crops as corn and soybeans. Overseas, and especially in the developing world, fluctuations in demand are further amplified by shifts in state agricultural subsidies for fertilizer use and by considerations related to current national balances of payments. As a result, exports to certain nations can exhibit dramatic swings from one year to the next, as in the 1994 export totals to China and India, which more than doubled and quadrupled respectively over the previous year (Table 2.2). Overall in 1994 overseas exports increased by 42 per cent, but they have just as easily shrunk by comparable amounts in other years. The volatility of foreign markets is drawn out further in Table 2.3, which presents the five-year sales' trend for the industry's five largest overseas customers. The main importers for Canadian product remain constant, although there is some movement in their ranking. On the other hand, the actual amounts imported exhibit considerable variation from one year to the next.

By virtue of being closely tied to the state of agrarian prosperity, prices received for potash also show great variation. Table 2.4 provides yearly average quotations for standard grade potash shipped out of the port of Vancouver.[8] From the early 1970s through to the present, the evidence of considerable price volatility for KCl is obvious.

As previously noted, partial nationalization of the industry by the government of Saskatchewan coincided with a boom in primary commodity markets and prices throughout the 1970s and on into the early 1980s. This inflation in primary commodity prices led some analysts, as well as politicians of various

TABLE 2.3
Five largest overseas potash markets, 1990–4

Year	Tonnes	Change from previous year (%)
1990		
China	1,322,551	20.7
Japan	584,486	−7.8
Malaysia	539,935	74.3
South Korea	396,746	−6.8
Brazil	388,814	−30.0
1991		
China	1,366,177	3.2
Malaysia	493,392	−8.6
Japan	481,035	−17.7
South Korea	445,219	12.2
Brazil	294,228	24.3
1992		
China	641,035	−53.0
Japan	513,026	6.7
Malaysia	388,263	−21.4
South Korea	338,869	−23.9
Brazil	293,373	−0.3
1993		
China	673,880	5.1
Japan	488,438	−4.8
Malaysia	375,023	−3.4
South Korea	373,398	10.1
Brazil	330,390	12.6
1994		
China	1,722,384	155.6
Brazil	789,092	169.0
Japan	518,298	6.1
Malaysia	470,397	25.4
South Korea	373,872	0.1

Source: Calculated from *Canadian Minerals Yearbook*, various years.

stripes, to advocate wholesale province/nation building strategies anchored around the escalating prices of such goods. This could be effected either by means of enhanced ground rental charges as first attempted by the provincial NDP government in the early 1970s, through joint public–private ventures, or via outright state ownership of such assets. It was this latter path that, at least partially, prevailed in the case of potash. However, in the context of the market

TABLE 2.4
Potash price quotations, 1972–95
(Current $US per metric tonne; fob Vancouver;
Standard grade KCl)

Year	Average	Change (%)
1972	$ 34	–
1973	34	–
1974	60	76.4
1975	82	36.6
1976	55	–32.9
1977	51	–7.2
1978	57	11.7
1979	77	35.0
1980	117	51.9
1981	112	–4.2
1982	83	–25.8
1983	75	–9.6
1984	84	12.0
1985	84	–
1986	68	–19.0
1987	69	1.4
1988	87	26.0
1989	98	12.6
1990	94	–4.0
1991	108	14.8
1992	113	4.6
1993	110	–2.6
1994	108	–1.8
1995	115	6.4

Source: Michel Prud'homme, Department of Natural
Resources, Canada.

euphoria of the period, the historical instability of product markets for such staples was often forgotten, only to return again with a vengeance as the decade of the 1980s wore on.[9] Tracking a depressed agrarian economy, and with systemic overcapacity in the industry, potash prices plummeted throughout most of the 1980s. Product prices, as indicated in Table 2.4, only began to recover in 1991. From then until the present, nominal prices have stood exactly where they had been eleven years previously; in real terms of course, they had still not recovered the ground lost in the 1980s.

Contrary to conventional wisdom and notwithstanding these trends, production in the industry has generally remained profitable. This has remained the case with the exception of the two years, 1986–7, when export prices completely bottomed out. This paradox could only have come about as a result of the internal reorganization that was taking place within the industry and that constitutes the heart of this study. In this case, market trends pose more of puzzle than a solution for understanding developments in the industry. With this return to profitability, the stage was set for the reprivatization of the industry. In 1989, during the dying days of the Devine government, PCS, the industry's largest producer, was placed on the auction block and sold off to private investors.[10]

If potash can now be considered a mature staple, uranium mining is still in a state of ongoing and rapid development. For its part, uranium oxide (U_3O_8), or 'yellow cake' as it is more commonly known, is the raw material out of which nuclear fuel is processed.[11] As such, it constitutes the basic input for nuclear power generation. Currently there are four uranium mine/mill operations in Canada, one in Ontario and three in Saskatchewan, where the world's richest known ore bodies associated with the Proterozoic unconformities are to be found in the Athabasca Basin in the north of the province.

Although the existence of uranium further north in the Lake Athabasca region had been known since 1936, development remained slow and was restricted to one location prior to the 1970s boom in exploration and mine development. The first commercial activity took place in the Beaverlodge area of Lake Athabasca where the federal crown corporation, Eldorado, sunk a shaft in 1949.[12] This quickly led to the establishment of Uranium City as the centre of mining activity for the next thirty years. Then, between 1968 and 1975, major new discoveries of richer ore bodies were made in what would later become the three currently operating sights of Rabbit, Cluff, and Key Lake mines. These finds overlapped with a resumption of U.S. imports of Canadian uranium and with a surge in the international price of uranium in the early 1970s aftermath of the oil crisis. In order to capitalize on what had become something of a uranium boom, the provincial NDP government created the Saskatchewan Mining and Development Corporation (SMDC) in 1974. Although the mandate of the new corporation was to explore for and mine all mineral commodities with the exceptions of potash, which was soon to be covered by the Potash Corporation, and sodium sulphate, in practice SMDC concentrated solely on uranium. Within this sphere the new crown corporation became a major proponent of the nuclear industry as well as the government's main vehicle for pursuing a politically hot development strategy.[13]

SMDC's mandate was further structured through the government's Crown Equity Participation Program. In accordance with this 1975 measure, companies acquiring provincial mineral dispositions were required to offer a 50 per cent share in any undertakings exceeding $10,000 to the new provincial crown company.[14] In this way SMDC was quickly twinned with such entities as Uranerz, the subsidiary of a German utility firm, with Amok, the French uranium consortium, and with the federal Eldorado Nuclear Ltd.[15] This provided SMDC and the provincial government with a major stake in the new uranium projects that were taking place at both Key and Cluff Lakes. By 1985 then, with assets of $898.2 million, SMDC had become the fourth largest producer of uranium in the world and had acquired 40 per cent of the known uranium reserves in the province.[16] In uranium, provincial state participation assumed the form of joint partnerships with private or mixed private/public foreign mining capital, as opposed to the strategy of partial nationalization employed in the case of potash. The objectives of state participation did, however, remain the same; assumption of a larger share of the economic rents associated with mineral production became the paramount means for costing out a social democratic provincial welfare state.[17] Ultimately, profitable accumulation lay behind such calculations.

Despite further ups and downs in product price, uranium development has moved ahead at a rapid pace in the province. Currently in the development stage are another three Saskatchewan properties, while three other mines in the province are in various phases of the environmental review process. As with potash, the industry was returned to the private sector (denationalized) with the merger of Eldorado and SMDC to form Cameco Corp. in 1988. Although Cameco began as a jointly owned crown corporation (61.5 per cent provincial and 38.5 per cent federal ownership), it has been sold off in stages, until today when all but residual shares reside in the private sector. Meanwhile the ongoing level of activity associated with the province's uranium boom makes Canada the world's largest producer of yellow cake. As indicated in Table 2.5, the country's share of 1993 production was 27 per cent of the world total. Of this, 93 per cent was produced at the Saskatchewan mine sites. Canada's share, however, was an actual decrease over previous years, when uranium production was effectively isolated into eastern and western geopolitical spheres of power. Canadian producers had been responsible for up to one-third of world production in the late 1980s. Unlike the situation facing potash producers, where former Soviet Union production has collapsed, the Canadian share of uranium exports is now somewhat diminished owing mainly to the entrance of the FSU onto the world market, and to a lesser degree, to the lifting of sanctions against South Africa.[18]

Although the domestic market represents a more important source of sales for Canadian producers than is the case for potash, it is still comparatively small. Only 15 to 20 per cent of domestically produced uranium concentrate is sold on the Canadian market, largely to Ontario Hydro, which owns all but two of the commercial reactors in service.[19] Total sales are dominated by the American market, which takes from less than half, to over 70 per cent of Canadian

TABLE 2.5
Production of uranium in concentrates by selected major producing countries, 1988–93
(tonnes of contained uranium)

	1988	1989	1990	1991	1992	1993
Canada	12,470	11,350	8,780	8,200	9,340	9,190
Russia	–	–	–	–	(in Other)	2,700
Kazakhstan	–	–	–	–	(")	2,700
Uzbekistan	–	–	–	–	(")	2,700
China	–	–	–	–	(")	950
United States	5,190	5,320	3,420	3,060	1,860	1,290
South Africa	3,850	2,950	2,530	1,710	1,670	1,710
Namibia	3,600	3,100	3,210	2,450	1,680	1,670
Australia	3,530	3,660	3,530	3,780	2,330	2,270
Niger	2,970	2,990	2,830	2,960	2,970	2,910
France	3,390	3,240	2,830	2,480	2,150	1,710
Gabon	930	850	710	690	540	550
Other	910	940	3,800	2,250	12,600	2,770
Total	36,840	34,400	31,640	27,580	35,140	33,120

Source: *Canadian Minerals Yearbook*, 1994.

production in any given year (Table 2.6). The other market of note, as shown in Table 2.6, is Japan.

The growth of world supply capacity and inventory is once again reflected in the prices received by Canadian producers. As indicated in Table 2.7, prices peaked in the resource boom of the late 1970s and early 1980s. This coincided with the sizeable expansion of the industry into the Athabasca Basin. Thereafter, uranium prices have shown a steady propensity to decline. The figures provided in Table 2.7 record the average yearly price on long-term export contracts, which is how most Canadian uranium is marketed. A certain amount can be sold on the international spot market, but in recent years, this has been insignificant, owing to the depressed state of current international prices. The longer term decline in contract prices, on the other hand, has been brought about by the completion of older, higher priced contracts with the mines at Elliot Lake, Ontario and with the entrance of the newer, cheaper Saskatchewan product onto the scene. As in the case of potash, and despite the commodity price profile of Table 2.7, the industry remains profitable, indeed even bullish. To account for the anomalies of overproduction, high inventory, and sustained profitability, it is again necessary to leave the international market place behind and probe closer to home, into the production sites themselves.

TABLE 2.6
Exports of Canadian uranium by destination, 1988–93 (tonnes of contained uranium)

Final destination	1988	1989	1990	1991	1992	1993
Argentina	–	–	–	19	20	29
Belgium	153	190	–	–	–	–
Finland	151	71	83	–	–	–
France	964	696	799	822	111	461
Germany	806	615	220	459	534	665
Indonesia	–	1	–	–	–	–
Italy	–	46	–	–	–	–
Japan	717	1,729	2,005	399	2,328	523
South Korea	874	635	339	215	104	715
Spain	100	97	–	–	–	–
Sweden	783	497	285	91	170	–
United Kingdom	1,204	871	882	498	19	–
United States	4,682	3,950	4,035	5,307	4,032	6,291
Total	10,434	9,398	8,648	7,810	7,318	8,684

Source: *Canadian Minerals Yearbook*, 1994.

First, however, to put this in perspective, the part of the mining industry considered in this book, potash and uranium, accounted for a value of close to $2 billion in 1994.[20] This rendered uranium the sixth most important metal, as measured by value of production, in Canada. Potash represented the most important non-metal commodity currently mined in the country. Although these figures are significant for a national political economy, they are magnified many times over in a small provincial economy. Thus, in any given year, potash and uranium production, taken together, vie with agriculture and manufacturing production in their contribution to Saskatchewan's provincial GDP.[21] Using other metrics, the total value of production in the provincial potash and uranium mining industries can be grouped alongside such other national industries as tobacco manufacturing, automotive engine and engine parts production, and the household furniture industry, all of which account for similar total values of output.[22] And last, a final telling comparison: total potash and uranium production in Saskatchewan accounts for almost twice the value of output as Quebec's automotive industry, which in itself is of no small accord.[23]

Although these remain abstract comparisons, intended only to provide an idea of the provincial and national significance of these industries, the most direct impacts are registered on those who work within them. It is to the industry labour forces that the study now turns.

TABLE 2.7
Canadian uranium export prices, 1974–94

| Year | Average export prices (C$kg/U) | | Spot sales portion of deliveries (%) |
	Current dollars	Constant 1994 dollars	
1974	39	109	n.r.
1975	52	133	n.r.
1976	104	245	n.r.
1977	110	243	n.r.
1978	125	261	n.r.
1979	130	247	n.r.
1980	135	231	n.r.
1981	110	170	1
1982	113	161	1.5
1983	98	133	10
1984	90	118	26
1985	91	117	20
1986	89	111	21
1987	79	94	35
1988	79	90	13
1989	74	81	<1
1990	71	75	<1
1991	61	63	<2
1992	59	60	<1
1993	50	50	<1
1994	51	51	<1

Source: *Canadian Minerals Yearbook*, 1994.

The Labour Markets

As is the case with other sectors of the mining industry, the potash and uranium industries of Canada are highly capital intensive undertakings. All told, uranium mining employed just over 1300 workers in 1994.[24] This included 564 employees at one Ontario mine and 756 workers at three Saskatchewan operations. Potash production employed over 3800 at two New Brunswick mines and ten Saskatchewan sites.

The workforce profile in potash reflects its status as a mature industry. Typically, an operation will employ between 200 and 300 workers assigned to underground mine and surface mill operations, mine and mill maintenance functions, and load out. Until the early 1980s, these numbers would have been considerably higher, but a decade and a half of ongoing mechanization in underground operations and computerization of mill and load out functions has taken its toll on employment. All of the Saskatchewan mines have witnessed significant levels of downsizing over this period. As a result, the workforce that remains is comparatively senior. In the four potash mines that are covered in this study, with a total operations and maintenance employment of 845, the average length of *job* seniority in our sample was eleven years.[25] This is the length of time that individuals have been in their current job positions. Average length of employment with the respective companies, meanwhile is sixteen and one-half years, with the workforce evenly spread in the thirty-five to fifty-five-year-old range.

These demographics lend a certain character to this workforce. First, it is almost exclusively male (98 per cent). Such was not always the case, at least not to the extent that one finds today. Female participation was greater throughout the industry's formative years and up until the recessionary 1980s, when exceedingly high rates of turnover characterized its workforce. Company managements made quite explicit appeals for women workers in this period, not out of commitment to the principles of employment equity, but solely in a bid to stabilize employment relations. However, as in other industries, the female component of the workforce generally had less seniority than its male counterpart.[26] When the downsizing of the 1980s took effect, women workers were the first casualties. In the interim they have all but disappeared from the mines and mills of the industry.

There is also little employment of visible minorities in the industry. In the sample that we worked with in the potash industry, 4.2 per cent of the respondents claimed First Nations ancestry. Although there is no firm data on the subject, there is little reason to suspect that this participation rate would have been greater in the past than it is currently.

A large proportion of the current potash workforce comes from rural backgrounds. When queried, 42.4 per cent of the industry sample indicated that their father's main occupation had been in agriculture while they were growing up (Table 2.8). Next most frequent were fathers who had been employed in the construction industry, followed by those who had been miners themselves. A majority indicated that their mothers had been homemakers (56.4 per cent), while 16 per cent reported that their mother's main occupation was farming, and 14 per cent had been brought up in homes where the mother was employed in the private service sector. These are occupational data, and it is not possible to rigorously transpose them into class categories. Nonetheless, they are suggestive. The largest proportion of the potash workforce sample hail from agrarian backgrounds, presumably from families who often found themselves in the position of owner/occupiers of the land that they worked.

TABLE 2.8
Potash workforce: father's and mother's occupational background
(%)

Occupation	Father	Mother
Agriculture	42.4	15.9
Mining	10.2	0.4
Other primary	0.8	–
Manufacturing	3.8	1.9
Construction	14.0	–
Transportation	6.4	–
Services: non-government	9.8	14.4
Services: public	8.0	9.8
Homemaker	0.4	56.4
Other	4.2	1.1

$N = 264$.

When we examine the prior work histories of this labour force, however, the agrarian experience begins to lose some of its salience. Most, 82 per cent, have occupied other full-time jobs prior to entering the mining labour force, but just over one-third (35.2 per cent) have had previous work experience in mining of one sort or another. This does not vary a great deal by occupational category; perhaps somewhat unexpectedly, skilled journey maintenance workers were only marginally more likely to have had previous experience in the mining industry than were mine and mill operators. Apart from mining, and calculated separately, the most common previous employments for the potash sample as a whole were found to be in the construction industry, manufacturing, private services, agriculture, and transportation, in that order (Table 2.9). At least some of today's workers were hired on as construction crew in the original development of the industry in the 1960s. When that phase had ended, they applied for and received jobs in the mining industry proper.

When it came to being hired, knowing someone already in employment at a mine site was advantageous for prospective job seekers. For 45 per cent of the sample, this was the most important means for obtaining employment in the industry, as compared with 39 per cent who indicated that they had simply submitted a general job application prior to being hired. Although results do vary from mine to mine, overall, the most important way of getting taken on was through already having a friend, family member, or acquaintance in employment, who could pass on word of openings or vouch for one's good character to a supervisor. The operation of impersonal labour markets were thereby modified in several ways by the sociological realities of place or location. Workers

TABLE 2.9
Previous job histories of potash workforce (%)

Occupation	
Agriculture	9.4
Mining	20.5
Other primary	1.5
Manufacturing	14.1
Construction	27.5
Transportation	9.0
Private service sector	14.1
Public service sector	3.0
Other	0.4

$N = 264$.

on the original construction and mine development crews sometimes stayed on, while the social networks of rural communities also served as major recruitment channels for the new industry.

Obtaining a sense of educational background for this workforce was more difficult, for the simple reason that the categories which are normally used to report educational attainment are not necessarily contiguous.[27] For example, it is possible, and indeed common, to find workers who have not finished high school, but who have completed a 'ticket' (that is, apprenticeship training) at a community college. As a result, college graduation does not necessarily entail high school completion. Because this is quite a common occurrence with practicing tradespeople, a better indicator of educational background is to be found by examining workers in production, who, with few exceptions, have not undertaken trades training. When this was done, it was found that the largest group, 45.5 per cent, was composed of high school graduates, followed by 22 per cent who had some high school education. Considering higher education as a whole, 23 per cent of the potash subset had completed some technical school or university education, or were graduates of community college apprenticeship programs, although not currently employed in a trades capacity.

The labour market in the uranium industry differs from what has just been described in several important ways. To begin with, under the Surface Lease Agreements entered into between companies in the industry and the government of Saskatchewan, the uranium companies have undertaken to hire both northern residents, and northern residents of First Nations ancestry.[28] The stated objectives have aimed for workforces that are minimally 50 per cent northern.[29] Thus hiring policy has constituted a large component of the 'northern development contribution' that has assumed a major role in the legitimation of a politically sensitive industry.

At Key Lake, which is the largest uranium mining operation in the province, the company's own data indicated that 42 per cent of a total workforce of 399 employees were of northern residential status.[30] This figure, however, must be treated with caution. It refers to workers who, at some time, have resided for ten years in the north, regardless of their current domicile.[31] In fact once they start with the company, many workers move to the south.[32] With the decommissioning of mining activities set to commence within the next year or so at Key Lake, it seems highly unlikely that the company will reach its own target of northern employment levels before the site is mined out. Aboriginal employment at Key Lake, at the same time, stood at 37 per cent of the workforce, or 147 employees.[33] These numbers were closely reflected in the interviews that were conducted of Key Lake workers, where Aboriginal and Metis workers constituted 35 per cent of the sample. Also significant was the higher proportion of female workers in the production end at this operation. Women workers constituted 13 per cent of the overall sample drawn at the Key Lake mine. Of this total, 75 per cent were of Aboriginal or Metis ancestry.

Although northern uranium mining predates potash development with the establishment of Uranium City on Lake Athabasca in the early 1950s, in its modern guise of what will be termed radically spatialized industrial relations, it dates back less than twenty years.[34] This is also reflected in the nature of the workforce in the industry. Here, on average, workers have occupied their current job positions for seven years, and have worked for Cameco for an average of ten years. Interestingly though, the modal figure for length of time in current position at Key Lake was only one year. This compared with a mode in the potash industry of fourteen years and is indicative of higher levels of workforce turnaround at the Key Lake site. The data on reported quit rates also reflect the fluid labour market conditions at Key Lake. When asked if they had personally known any workers who had voluntarily quit their jobs with the company in the last two years, 87 per cent of the Key Lake sample responded affirmatively. On average, 8.5 quits were recounted among the members of the sample, compared with 78 per cent of the potash interviewees who cited a mean of 3.8 quits over the previous two years.

The existence of the Surface Lease Agreements, and the employment equity programs that they entail, makes it difficult to refer to a single labour market at Key Lake. A north/south divide in the workforce is evident both with respect to the very different work histories that have been brought to the employment relation with the company and with the transposition of these biographies onto the occupational structure at the mine. With respect to the latter, Aboriginal and

TABLE 2.10
Uranium workforce (Key Lake), occupation by race[a]

Race		Mine operations	Mine trades	Mill operations	Mill trades	Row total
Aboriginal	N	11	3	6	2	22
	%	50.0	13.6	27.3	9.1	35.5
	%	61.1	25.0	33.3	14.3	
Non-Aboriginal	N	7	9	12	12	40
	%	17.5	22.5	30.0	30.0	64.5
	%	38.9	75.0	66.7	85.7	
Column	N	18	12	18	14	62
totals	%	29.0	19.4	29.0	22.6	100.0

Cramer's $V = .37$
$p = .03$
[a]The first numbers in each cell are the raw freqencies, followed by row percentages and then column percentages.

Metis workers were clearly over-represented in mining operations and under-represented in the skilled trades on site. Half (50 per cent) of the indigenous workers surveyed were working in the mining department, usually as heavy equipment operators in the open pit, while another 27 per cent worked in the mill as operatives (Table 2.10). Again, although 42 per cent of the total sample were located in journey trade positions (mine, 19.4 per cent and mill, 22.6 per cent), for Aboriginal and Metis workers only 23 per cent were so occupied. Examined from a slightly different angle, Aboriginal workers in our sample constituted 35.5 per cent of the total workforce, but 61 per cent of the mine pit crews as compared with one-quarter of the skilled mine trades workers and an even smaller 14 per cent of the mill trades people. Notably, throughout the whole of the operation, there was not one supervisor of Aboriginal or Metis ancestry. As Table 2.10 shows there has clearly been a strong association between race and position within the occupational division of labour at Key Lake (Cramer's $V = .37$). This overlapping of race and occupation made for a far less homogeneous workforce than was found at the other sites in this study.

Often union seniority rules are implicated in the production of such dualism. That is, members of the favoured group are hired first as a result of existing opportunity structures. As a result, they hold the highest levels of employment seniority. A focus on job control unionism and the accompanying rules render it exceedingly difficult to challenge the hold that such groups then have on pre-ferred jobs. This was certainly one perception that existed at the mine site, even

TABLE 2.11
Uranium workforce: father's and mother's occupational background, Aboriginal and non-Aboriginal workers (%)

Occupation	Aboriginal (N = 22)		Non-Aboriginal (N = 41)	
	Father	Mother	Father	Mother
Agriculture	13.6	–	51.2	4.9
Mining	13.6	–	12.2	–
Other primary	27.3	–	–	–
Manufacturing	4.5	4.5	–	2.4
Construction	13.6	–	17.1	–
Transportation	–	–	2.4	–
Private service sector	9.1	27.3	9.8	29.3
Public service sector	13.6	18.2	7.3	7.3
Homemaker	–	50.0	–	53.7
Other	4.5	–	–	2.4

though this type of explanation for the existing occupational structure does not account very well for the facts of work at Key Lake. With respect to seniority levels, the differences between Aboriginal and non-Aboriginal workers were actually quite slight in our sample. Average employment seniority at the mine was 10 years. The corresponding means for Aboriginal and Metis employees were 9.6 years and for non-Aboriginal workers, 10.2 years. Additionally, there was slightly more than a one unit job class difference between the average positions filled by Aboriginal workers and those staffed by non-Aboriginal workers, in favour of the latter. Exactly the same difference showed up in the mine department where Aboriginal workers were most frequently employed as equipment operators in the pit. Here the average job classification for these workers was 7.3 on the company's job classification grid, while for non-Aboriginal workers it was 8.3.

Differences in occupational backgrounds were also a clear feature of labour markets in this industry. As with their counterparts in potash, agrarian backgrounds figured prominently in the biographies of uranium workers from the south. Overall, 51 per cent cited agricultural employments as their father's chief occupation (Table 2.11). Construction sector employment and work in mining also figured in intergenerational employment patterns as was the case in the potash workforce, although agrarian backgrounds far outweighed the other pursuits. For their mothers, home-making was once again given as the most frequent labour, although almost 30 per cent of parental households found the mother working in the private service sector.

TABLE 2.12
Previous job histories of uranium workforce, Aboriginal and
non-Aboriginal workers (%)

	Aboriginal	Non-Aboriginal
Agriculture	–	2.1
Mining	30.6	26.1
Other primary	5.6	1.1
Manufacturing	5.6	9.8
Construction	27.8	33.7
Transportation	5.6	6.5
Private service sector	13.9	12.0
Public service sector	11.1	8.7
Other	–	–

For workers of Aboriginal descent the biographies were different. Here 27 per cent of the sample had fathers who worked in other primary undertakings, such as fishing and trapping. This was the most important economic activity for fathers, followed by work in the mining industry, construction, and employment in government services, which were all equally important. Exactly one-half of the First Nations workers indicated that their mothers had been exclusively homemakers, while 27 per cent had mothers with a history of employment in the private service sector.

Considerably higher proportions of uranium miners indicated previous work experience in the mining industry than was the case for the potash workforce (Table 2.12). Approximately one-third (35 per cent) of the potash sample had possessed previous job experience in mining, but in the case of uranium, over one-half (55 per cent) of the sample had entered the employment relation with previous mining experience. This may well be a function of the lack of other, comparable employment opportunities for those who live in the north. Thus, previous work at a uranium mine prior to migrating to Key Lake would not be an unusual occurrence. At least a few of the southern workers were previous victims of downsizing in the potash industry. With employment no longer available in this sector, personal decisions were made to seek work in the newly expanding uranium industry.

Apart from prior work in mining, previous job histories for Aboriginal workers were to be found in the construction industry, private service sector, and public service sector. Each of these employment categories was more important than previous work experience in the primary sector, indicating an intergenerational movement out of the latter akin to the movement of southern workers out of the agrarian sphere over the course of two generations. For workers from the south who were employed at Key Lake, previous occupational histories in the construction industry, private service sector, manufacturing, and public sector (in that order) loomed large.

The frequency of previous experience in the mining industry, for employees at Key Lake, may also account for the absence of reservations about working in this particular industry. Employment at Key Lake is notable in two respects. First, it entails a further, radical attenuation of the separation between working life and civil society than is normally found in capitalist society. This dimension of work at the company is discussed in greater detail in the next chapter under the notion of radically spatialized industrial relations. Second, the handling of sensitive materials (U and U_3O_8) and the issue of safe exposure levels is a daily aspect of the labour process in this industry. Given this, only 15 per cent of the sample indicated having reservations concerning employment in a fly in/fly out work camp situation, prior to commencing their jobs, and an even lower 7 per cent, expressed concerns about working with what some would define as a hazardous product, U_3O_8.

For workers, recruitment into employment at Key Lake differed little from what was found in the potash industry. Having a friend or family member already in employment was cited as the most important factor in the recruitment process by 43 per cent of the Key Lake sample, followed by submission of an unsolicited job application by 35 per cent. Interestingly, having a contact in the company was considerably more important for southern miners than was the general job application (46 per cent as opposed to 33 per cent), although the opposite held true for Aboriginal workers, where unsolicited applications were attributed to 41 per cent of the hirings, and family or other social relationships were cited by 36 per cent as being the most important factor in hiring.

For company management, on the other hand, recruitment to an isolated fly-in site that respects affirmative action and equity guidelines poses challenges that are not faced in the potash, nor for that matter, in many other industries. Under a considerable amount of public scrutiny, Cameco has definitely been sensitive about its northern hiring mandate. To this end, the company has retained a northern hiring office, as well as a manager of Northern Affairs, Training, and Education, whose role has been to function as a liaison with northern communities. At one time, all job requisitions would be sent out to native communities throughout the north, but recently the company has had such a backlog of applications in its designated job bank that this has no longer been necessary. In line with our own data, the filling of entry-level operating positions from the north has not been problematical. On the other hand, skilled maintenance positions have only seldom been occupied by Aboriginal/northern populations. In 1993, the company entered into a multi-party training plan with the provincial and federal governments, with the intent of providing skills upgrading for northern residents. To date though, only a limited number of first-level apprenticeship courses have been offered anywhere in the north. After that, students have had to travel to the south for three more levels of training in such fields as mechanics, carpentry, and electrical apprenticing.[35] In many cases, this continues to constitute an effective obstacle for additional training and upgrading.

Aside from actual recruitment, the company has employed a number of other labour market-related strategies that pertain to the unique circumstances of hiring and working in a remote environment. Liaison with and participation in the scattered northern communities from which the company recruits have been judged to be important ancillary activities of management. Under this guise Cameco has offered a steady stream of site tours for northern high school students and community elders. Work placements for northern high school students have also been utilized to familiarize northern communities with the career possibilities that are offered through the company. This support work represents an extension of the employment equity program that has been part of the mandate received from the state. It has also been one possible means of countering turnover rates that reached as high as 80 per cent for northern workers in the first years of operation.[36] Given these two factors, above average rates of company resources have been going into labour market activities. However, recruitment is one thing and retention is another. Although company management has acknowledged that considerable amounts of family and community support are needed for the maintenance of radically spatialized work relationships such as those found at Cameco, it is less clear where managerial responsibilities lie in this unique equation.

In assessing previous educational backgrounds in the uranium industry, some differences are found from what was previewed in potash. In this instance, the largest proportion of the subsample, which includes mine and mill operators, indicated having received some high school education. Forty-seven per cent of the sample fell within this category, while 24 per cent had completed a high school education, and 21 per cent had some level of higher educational attainment.

From the standpoint of labour power, we should not assume that all labour market activity ceases with the commencement of employment at a mine. In the case of both the potash and uranium industries, a considerable number of workers (31 per cent of the potash sample and 33 per cent of the uranium) combined work at the mine with a range of other jobs. In the south, *second* jobs were most commonly part-time activities related to farming, followed by work such as freelancing in one's trade for the local construction industry, and employment

in a variety of private services. The latter illustrate a surprising diversity, ranging from personal-care product distribution through such companies as Amway,
to computer and income tax consulting services. For the uranium workforce,
composed of both northern and southern residents, *second* jobs in personal services and construction were the most important. These pursuits ranged from
carpentry and plumbing to wilderness guiding and logging. Intriguingly, and
perhaps related to the second-job phenomenon, almost one-third of the sample
in each industry evaluated their current wages as being just enough to get by on,
or as inadequate for present needs.

Up until this point, in the discussion of product and labour markets, it has
been permissible to remain at the general level of industry characteristics. The
commodities that are being produced, KCl and U_3O_8, are standard goods, produced for large export markets. The labour power that is used to produce them,
although skilled in certain regards, is not so uniquely qualified as to be able to
constitute an exclusive labour market. In beginning to inquire into the nature of
work relationships and the social organization of employment relations, however, it is necessary to become more specific. Neither the product, the labour
market, nor the production process itself, determines a singular mode of social
organization at these workplaces. Instead, each corporate entity differs with
respect to what are considered to be best practices and paradigms for arranging
the organization of the work-effort bargain and managing the workforce. An
appreciation of the variety in current managerial practices, real and imagined, is
a prerequisite for any analysis that purports to examine the labour force impacts
of the new industrial relations. A detailed overview of the different corporate
cultures of employment that make up this study is provided in the following two
chapters.

Summary

Two items stand out in this review of the product and labour markets that characterize the potash and uranium industries. With respect to labour markets, both
industries employ a first-generation mining workforce. By and large, these
workers did not hail from mining families, or mining communities. The occupational backgrounds of parents, and perhaps their class backgrounds as well, differed from the current generation of potash and uranium workers. As a result,
there are fewer historical or cultural traditions for these employees to draw
upon in their dealings with the companies than is the case in older mining
regions.

The second feature of note is the rupture in economic relationships that normally hold between product markets, prices, and profits. Despite overcapacity

and overproduction in the potash industry, and growing supply schedules coupled with flat markets in uranium, producers continue to enjoy golden conditions. Although commodity prices have only just recovered in nominal terms for potash producers, or continue to slide for uranium manufacturers, corporate fortunes have seemed immune to the short-term vagaries of the market place. In order to discover how this might be so, it is necessary to leave behind the ebb and flow of commodities and their exchange values and enter directly into the sphere of production. The following chapters discuss the different modes of organizing the work-effort bargain at the mining companies in greater detail.

3

Corporate Cultures of Employment I: Two Traditional Firms

Corporate Cultures and the Management of Workers

In one of his better-known passages, Marx writes that the sole difference between various historical modes of production amounts to 'The specific economic form, in which unpaid surplus-labour is pumped out of the direct producers.'[1] Just as important for our purposes is the coda that Marx adds to this: 'This does not prevent the same economic basis ... from showing infinite variations and gradations in appearance, which can be ascertained only by analysis of the empirically given circumstances.'[2] It is this qualification that is of interest here, for capital accumulation can itself support an astonishing variety of workplace relations, within the wage relationship. These variations in the organization of the wage relation are likely to reach beyond the realm of appearances alone, creating real life effects on producers and their relationships with corporate owners and managers. Especially within periods of social crisis and questioned hegemony, finding only one organizational format, or one unchallenged way of doing business, is unlikely. Instead, divergent strategies of organization within a system of accumulation are more likely to be encountered. Often these are presented as 'best practices,' which can be distilled into methods for the effective measurement of work effort and result, while being generalized into full blown paradigms that proffer efficient models of work organization for anxious managers.

Such paradigms and the practices that inhere to them are the subject matter here. In one of the first efforts at analyzing such a paradigm, Gramsci pointed out that the American system of production, Fordism, went beyond a defining technology, or a set of techniques, to actually embody a new culture, or way of looking at things; 'the new methods of work are *inseparable* from a specific mode *of living* and *of thinking* and *feeling life* [my emphasis]. One cannot have success in one field without tangible results in the other. In America rationalization of work and prohibition are undoubtedly connected. The enquiries conducted by the industrialists into the workers' private lives and the inspection services created by some firms to control the "morality" of their workers are necessities of the new methods of work.'[3] Fordism, as Gramsci suggests, stands for a new hegemony, 'born in the factory,' 'a skillful combination of force ... and persuasion,' that reaches far beyond in succeeding to make 'the whole life of the nation revolve around production.'[4] Hence, Fordism entails a whole culture, Americanism, which finds popular expression in work habits, norms of consumption, domestic/familial arrangements, and sexuality.[5]

For Gramsci, Fordism backs a societal culture, Americanism, and a specific culture of employment. It is both a means of organizing and obtaining work effort – *a labour process* – and a broader host of supporting practices, which constitute the *political apparatuses of production.* Within the corporation, such apparatuses are constituted by a host of practices – *production politics* – which include, but are not restricted to, prescribed forms and channels of interaction and communication, sanctioned modes of problem definition, enjoined goals, lines of authority, modes of representation, and methods of reward. When taken together these characteristics stamp an organization with a given identity and culture, as surely as the product that it manufactures.

This study includes four multinational mining companies at five different worksites as a way of initiating an analysis of two different cultures of employment – Fordism and post-Fordism. Selection of the companies was based upon a number of considerations, but most important, upon the range of practices that are exhibited in the organization and in the appropriation of productive effort. Additionally, each company is a major institutional entity in its product market. The Potash Corporation of Saskatchewan (PCS) and International Metals Corporation (IMC) are respectively the largest and second largest potash producers in the global market. Similarly, Cameco is the world's largest uranium concern. Size alone merits some attention, but, as indicated above it is not decisive. Just as important, each company offered up a range of practices that symbolized different cultures of employment. Here it is important to keep in mind the object of our analysis – alternative organizational structures and their impacts upon those who are subjected to them. The actual corporate entities, then, are intended as exemplars of divergent practices, as enumerated below.

The corporations that have been used as representations of differing paradigms in this study include, in the potash sector, the Potash Corporation of Saskatchewan (two divisions), Agrium Fertilizers, and Central Canada Potash; and in the uranium industry, Cameco Ltd. The presence of two PCS sites requires some additional elaboration, before going further into an ethnography of each organization.

PCS-Cory, introduced in the first chapter, provides the best example of radical downsizing that exists in the industry. At PCS-Allan, on the other hand, human resource management has followed a different, although thematically related tactic, by institutionalizing a seasonal employment regime. These represent two divisions of one company, where one has gone through considerable job loss while the other has entered into a more contingent relationship with its workforce. In both instances, a strategic use of labour and commodity markets has provided a prime lever through which to realize corporate labour policy, although in somewhat different ways. A comparison of the two sites could thereby prove instructive.

The remainder of this chapter introduces the two firms that fall towards the Fordist end of the spectrum, PCS and Cameco Ltd. The ethnographic narrative is resumed in Chapter 4, where two firms that are moving away from Fordist techniques for organizing the labour process, Central Canada Potash and Agrium, are examined in detail.

PCS Inc.: Fordism and Radically Marketized Industrial Relations

As explained in the last chapter, the Potash Corporation of Saskatchewan was created by an act of legislation in 1975, in the midst of ongoing court challenges by existing private producers to the introduction of a new provincial Reserve tax on the industry. Up to this point, 62 per cent of the industry had been American owned, 11 per cent was European based, a South African firm owned 9 per cent, and the remaining 18 per cent rested in Canadian corporate hands.[6] Within the next three years, four mines had been purchased outright by the government of Saskatchewan, along with a 60 per cent interest and manager/operator status in a fifth property, Saskterra Inc.[7] These acquisitions provided the new crown corporation with considerable industrial clout. By the early 1980s, PCS held 28 per cent of total capacity in the North American potash industry.

Privatization in 1989 has since led to another acquisition spree on the part of the company. The remaining 40 per cent of the Allan potash mine that was still held by Saskterra Ltd. was acquired in 1990, followed by the 1993 purchase of another large U.S. mining concern, the Potash Company of America and its properties in Saskatchewan and New Brunswick. This latter purchase added another two mining properties to the company's holdings.

There has also been considerable complementary diversification on the part of PCS management in the interim. First, in 1991, the American firm Florida Favorite Fertilizers was purchased with its assets in the southern United States. This gave PCS ownership of what is described as a 'full range fertilizer company, producing and selling blended, granulated and solution-mixed fertilizers in addition to farm chemicals.'[8] To this was added Texasgulf Inc., a major producer of phosphate chemicals, purchased in 1995 from the French oil, chemical, and pharmaceutical giant, Societé National Elf Aquitaine SA. This acquisition gave PCS a major presence in two of the three inputs that are required for all commercial fertilizer: potash and phosphate. The third ingredient was added to the company's repertoire in 1996, with the purchase of Arcadian Corporation, the largest nitrogen producer in the western hemisphere.[9]

This represents an interesting turnaround. Created by state charter out of the nationalization of several foreign-owned companies in the mid-1970s, PCS is now a multinational corporation in its own right. Capitalized at over $2 billion, the company's forty-three million shares are now traded on the exchanges of New York, Toronto, and Montreal.[10] In total, PCS currently controls over one-fifth (22 per cent) of world KCl capacity and is poised for further expansion into the growing markets of the Third World. It is in this context that the company has acquired mining concerns and port facilities in Eastern Canada to serve the growing Latin American markets, as well as branching out into phosphate production, of which it is now the number three producer in the world. Even now, offshore sales of potash represent 60 per cent of PCS's business, up from 45 per cent in 1981, with the remaining 40 per cent dedicated to the North American market.[11] If anything, this turn towards the global market place is likely to continue for the company, with its recently announced offer to purchase majority ownership of the German Kali und Salz AG, the holding company that controls the whole of the German potash industry. This deal will add substantial capacity to PCS's already impressive productive inventory, and bring its share of mineral reserves up to between 30 and 35 per cent of world capacity. This could translate into a 50 per cent plus share of world trade in potash.[12]

The changes in company status, outlined above, have had some interesting implications for production politics at this company. In a valuable dissertation, June Corman has plotted out the effects that partial nationalization in the industry had on employment relations, comparing PCS's policies with those of both its predecessors and other firms in the industry. In the main, Corman uncovered several small differences at the margins of the employment relationship and one major shift that came with nationalization. Among the minor changes in employment relations that followed in the wake of PCS's creation were the introduction of expedited grievance proceedings for individual disputes, increases in the number of worker health and safety representatives, the initiation of employee recovery programs, the posting of first line supervisory positions, and paid time off for local negotiators. Although PCS's political managers also saw fit to appoint one trade unionist from outside the industry to sit on the company's board of directors, a possibly more significant development occurred with the chartering of a Scandinavian-style work environment board for the company in 1981, late in the governing social democratic party's term of office.[13] As this tripartite body struggled mainly with defining the parameters of its mandate during the year and a half before being nullified by the incoming Conservative government, it is not possible to say where a fully functioning environmental work board may have led. Be that as it may and although each of the modifications signaled a desire for a more cooperative industrial relations climate, overall, Corman found no evidence of greater industrial democracy at PCS than elsewhere. As both Sass and Corman conclude, industrial democracy was neither willed by management, nor envisaged by the social democratic politicians who had overseen the creation of PCS.[14] On the other hand, PCS's mandate did diverge from other companies in one significant aspect – the avoidance of unemployment or inventory control lay-offs at the company. This directive was formally announced by management as a 'no lay-offs policy' in 1981.[15] As Corman sums up: 'PCS's conduct in the short term deviated from the approach that a private producer would have followed in one important respect. PCS minimized lay-offs during periods of oversupply by transferring production workers to construction jobs.'[16]

Although public ownership was a necessary condition for a regime of full employment at PCS, it was not a sufficient condition, as the election of a Conservative government in 1982 was to show. Immediately upon taking office, the new government announced an end to the lay-off avoidance policies of its predecessors and followed this up with a four-month shutdown of mining activity at the company.[17] As Corman, and Laux and Molot demonstrate, full employment in the nationalized sector of the potash industry was a conscious political undertaking that was made by the social democratic government of the day. When this government was gone, so too was a defining feature of the industrial relations scene that characterized PCS in the early years. Just as a societal commitment to full employment would expire in the early 1980s, 'internal corporate Keynesianism' would be subject to the same fate at approximately the same time.

From 1982, and prior to the privatization of PCS, different production politics determinative of a new culture of employment took hold at the company's properties. It has been composed of both formal human resource management initiatives and a broader shift in corporate strategies that have had profound impacts upon the labour force. Thus, in a manner akin to the observations of the 1968 Donovan Commission into industrial relations in Britain, two systems of workplace control might be hypothesized as existing at PCS: a formal system that has been authored by HR managers, supervisors, trainers, and union officials; and a second unwritten discipline enforced by the market, which I will term radically marketized industrial relations.

Three elements enter into the formal relations of ruling at the company.[18] They include the systematic use of detailed training modules throughout the operation; the adoption of a Loss Control program as the strategic centre around which management of the labour process is geared; and a cooperative wage study protocol for the classification of all unionized jobs on site. Significantly, the two managerial initiatives, modular training techniques and Loss Control, were introduced to PCS after 1982 in the years of transition to private corporate status. The cooperative wage study program (CWS) was introduced one year earlier in 1981, following a two-week strike in support of its adoption.

Operations at PCS are now thoroughly 'modularized.' For every piece of equipment on site, no matter how simple or complex, there is a training module that instructs workers in the proper procedures entailed in its utilization.[19] To take an example, one module is used to instruct novice miners on available underground communications systems, including the use of telephones, cap lamp signals, and written messages. At the outset the performance objectives are set forth in each module, for example, being able to 'describe and perform proper procedures for operating a telephone' in an underground environment.[20] Each procedure is then broken down into its component operations with precise instructions that the employee is expected to know. Thus, phones should not be tampered with; the emergency number should be committed to memory; and the phone should be operated in the same manner as a home telephone. Sticking with the underground communications module for a moment longer, it is further specified that written messages 'must be understood.' They are to be legible and must avoid the use of profanities.[21]

Such modules are written by designated training foremen. Typically, individual workers are given some time off during the shift to 'study' several such 'mods.' This is followed by an open-book examination which is written and sometimes rewritten(!) until the module is 'passed.' Employees are then expected to validate the experience by signing the appropriate company form, attesting to the 'fact' that they have received training on specified pieces of equipment. It is not uncommon, however, for workers to refuse to lend their legitimacy to this exercise by refusing to go along with this last measure.

Clearly, the company takes the training modules seriously. They are, for example, highlighted in PCS's Annual Report, significantly in the section dealing with the company's safety record where it is noted that 'safety is built into the modular training programs ... Proper procedures are developed for each task of a job.'[22] On average, members of the Allan sample that were interviewed, reported taking 24.5 modules each, while at Cory the corresponding figure was 21 modules per respondent. These are much higher numbers than are reported for any of the other sites in the study.

Perceptions of the workforce, however, are somewhat at variance with the official discourse surrounding the training modules. Most frequently this training is identified as a simple requirement of the job, although at the Allan mine almost one-fifth of those interviewed related this system to a shifting of liability for accidents over to the worker. Here workers suggested that attesting to receiving and 'graduating' from company training programs could relieve the company of further responsibility. For this reason some workers withhold their consent and their signatures from company training records. Only small numbers at either Cory or Allan saw any relation between modular training and one commonly ascribed function of such education, namely, learning how to operate new pieces of equipment or gaining familiarity with new technology.

If anything, an added emphasis has been given to modular training at PCS with the adoption of a Loss Control system as the guiding principle of workplace management at company sites in 1987. The program is in effect at Cory, Allan, and the other company divisions. It was purchased from the International Loss Control Institute of Loganville, Georgia, and featured an initial five-day management training seminar, conducted by the institute.[23]

Loss Control is about the minimization of risk, a topic that is perceived as a 'science' in its own right. Accordingly, 'at its best, risk management can be one of the most profitable activities that an organization can pursue.'[24] Risk and loss are defined here as involving all facets of an operation: the people (human capital) and industrial accidents; site property and losses through theft, et cetera; and loss of potential income through productive inefficiencies that cause accidents and other stoppages. Loss is problematized as being an outcome of inadequate managerial control. As identified in the Loss Control training manual, 'this first domino in the sequence of events that could lead to a loss is the "lack of control" by management,' where control is defined as the authority to plan, organize, and lead.[25] The program then addresses 'job indoctrination' whereby 'each employee has received the proper job indoctrination prior to the start of his work activity.' In turn, this assumes the development of standard job procedures and training instruction, as well as measures of program compliance to which all are to be held accountable.[26] In the development of standard work procedures, job analysis is to be utilized. 'Every job can be broken down into the sequence of steps that take place to do it. There is usually a particular order of steps that is best to do the job most effectively, and it is the orderly sequence of steps that will eventually become the basis for the proper job procedure ... Experience has proven that most jobs will break down into ten to fifteen key steps.'[27]

Systematic job observation by management is a large part of Loss Control at two levels. First, it is utilized to develop proper job analyses through observation and benchmarking of best practices. And second, it is used to measure conformity with those practices. As proffered by the institute, 'The Planned Job Observation is a proven technique which enables a supervisor to know whether or not a worker is performing all aspects of a specific job with maximum efficiency. Maximum job efficiency means greater and safer production at lower costs.'[28]

As one goes further into Loss Control, there is more interesting slippage. Commencing under the guise of an accident prevention program, Loss Control comes to assume a preoccupation with property protection and income maximization. Safety becomes conflated with efficiency, which in turn is directly related to costs of production and profitability. As Alan Hall found in his analysis of health, safety, and the labour process at Inco, Loss Control has three key objectives: exerting greater managerial control over worker activity; reinforcing workers' commitments to individual personal responsibility for safety; and getting workers to acquiesce to the new 'economic realities' that tie health and safety issues to new technologies, production, and profitability.[29] Those who have the impression of 'old wine in new bottles' have not gone far astray. Despite some new wrinkles, including an emphasis on accident reporting and analysis, and the highly visible rating of departments for their conformity with the program, Loss Control retains all of the essential features of scientific management. This includes the use of systematic job observation dedicated to the production of planned and highly detailed work procedures. For all practical purposes then, Loss Control is a contemporary manifestation of scientific management.

The third and final component of formal industrial relations at PCS is the cooperative wage study method for evaluating jobs and remunerating work. Unlike modular training and Loss Control, CWS is an *negotiated* program. However, even though it was introduced at the behest of the union, this does not imply any incompatibility with the managerial initiatives that have already been described. If anything, CWS was intended to dovetail with, and actually promoted, the development of work situations premised upon a detailed division of labour. What CWS does do is remove some of the determinations regarding levels of payment and promotion from unitary managerial control. In other words, CWS is a method of job measurement that allows for a co-determination over the final results. Devised by the steelworkers in 1946, CWS was adopted at PCS following a two-week strike in 1981 at the Allan mine site.[30] Under CWS, management retains the right to define, design, and delimit jobs. The job, as constructed by management, is then subject to evaluation by a company and union CWS committee, where it receives a rating and a commensurate rate of pay. Failing agreement, classification disputes are referred first to a nominated company and union referee and then, if necessary, to an external arbitrator for final resolution.

In these matters CWS brings a form of industrial governance to the workplace. As in any polity, there are certain rules that are brought into play and that are recognized by the parties to be legitimate. With CWS there are two principal rules. First, it is jobs, not individuals who are being evaluated. This is to be done without regard for pre-existing wage rates.[31] Second, CWS implies a 'full knowledge regarding the functions of each job and its requirements through a job description ... The importance of adequate job descriptions cannot be overstressed.'[32] It is the job description, including titles, functions, forces of production used, and mandated procedures, that forms the basis of the subsequent evaluation.

Job ratings are constituted by composite weights given to the twelve variables that enter into every evaluation. Those items that are considered and scored include, amongst other things, any pre-employment and employment training requirements for the job, mental and manual skills that are exercised on the job, responsibilities for equipment and people, expended mental and physical efforts, and the presence of potential hazards in the work environment. As it is the job rather than the individual being evaluated, CWS lies at the heart of what has become known as job control unionism.[33]

Prior to the advent of CWS at the Allan mine, there were eight different job classes and payment rates.[34] A year later, this had blossomed into twenty job classes, replete with forty-five different occupational titles. Between each class stood a $0.28 per hour differential.[35] An updated version of this classification schedule is provided in Table 3.1. This definition of specific job classes in turn allows for the mapping of lines of progression within each department. Figure 3.1 furnishes one example of this for the underground mining department at Allan. Such lines of progression are constituted as the basis for the job bidding and bumping rights that regulate the internal labour market at the site of production.

Two points bear reiterating in regard to CWS. First, at PCS, as elsewhere, CWS was proposed by and adopted at the insistence of the union, as a means for insuring greater control and fairness over wage levels.[36] And indeed, the union did exercise a thoroughgoing control over the job evaluations, which were actually conducted prior to the strike. Almost all employee classifications received a substantial wage boost, in some cases of up to $2 or $3 per hour. On the whole, wage costs rose by approximately 22 per cent between the 1981 and 1982 collective agreements, of which over 16 per cent was attributable to the adoption of CWS.[37] Only in retrospect did the company realize what it had been saddled with and, when it did, management balked at the adoption of the plan. This precipitated the two-week strike that led to the final adoption of the scheme.[38]

TABLE 3.1
Job classes and wage rates, PCS-Allan, 1993

Job class	Wage rate ($ per hour) 1 May 1993
20	21.53
19	21.12
18	20.71
17	20.30
16	19.89
15	19.48
14	19.07
13	18.66
12	18.25
11	17.84
10	17.43
9	17.02
8	16.61
7	16.20
6	15.79
5	15.38
4	14.97
3	14.56
2	14.15
1	13.74

Source: Potash Corporation of Saskatchewan Inc., Allan Division and United Steelworkers of America, Local 7689, 1992–4 Agreement.

Second, it must be acknowledged that there is nothing in CWS that militates against the company's initiatives of modular training and Loss Control. Although CWS does allow for union voice over the rating of jobs, this is only required when a detailed division of labour is in effect. Ergo, CWS is best conceptualized as a pluralistic counterpart to the technical division of labour at PCS. Some analysts have linked CSW with a broadening of job categories and responsibilities, while others have suggested that CWS is the epitome of job control unionism, a form of regulation that militates against flexibility at the expense of labour power in the labour process.[39] All of the evidence that I have come across in interviews with mine managers and union officials supports the latter view. In short, the managers who are saddled with CWS view it as an obstacle to more flexible labour power deployment and would like to be rid of

FIGURE 3.1
Lines of progression, Mine Operating Department, PCS-Allan

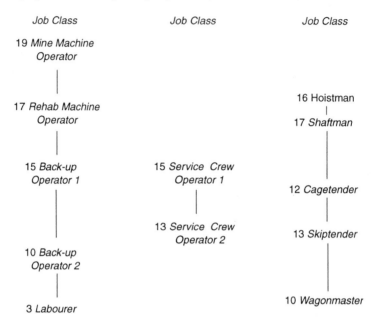

Source: PCS Inc. Allan Division and USWA Local 7689, 1992–4 Agreement.

it. On the other hand, one mine manager who does not have CWS at the Agrium worksite, wouldn't want to touch it.[40]

Modular training, Loss Control, and CWS constitute the mainstays of documented labour force control at PCS. But they do not provide a complete repertoire of relevant managerial practices at these properties. For that, the influence of the product market, and the issue of inventories and excess capacities must also be factored back in. The company has come to use these variables in a highly strategic fashion in its dealings with the workforce.

In 1994 PCS posted a record year with sales of 5.2 million tonnes of KCl, or a 36 per cent increase over the previous year.[41] Even at this volume, however, the company was only just taking advantage of less than half (48 per cent) of its productive capacity.[42] To some this might spell crisis, but it is not necessarily the case for PCS. As the corporate profile explains, 'Rather than using all its production capability, PCS matches production to sales demand to stabilize prices; that is its ongoing strategy.'[43] Elaborating on this point further, a recent annual report clearly states: 'Market-driven, rather than production-driven, PCS remains convinced that price, not volume, is the key to its financial success.'[44] The company is, in practice, willing to cede a certain share of the market in exchange for commanding a higher price for the volume that it does sell. Higher market profits are realized by producing less and charging more, or as managers cutely phrase it, PCS is no longer 'the dog that is being wagged by the tail.'[45] It is precisely this orientation that I have tried to capture in the notion of radically marketized industrial relations.

Such relations are manifested in a number of ways at PCS. Beginning in 1982, the commitment to employment stability was annulled. In that year, initiatives on the part of PCS to solidify long-term delivery contracts were abandoned by the newly elected Conservative government.[46] At the same time, lengthy, seasonal lay-offs were instituted at the Cory property, culminating in the adoption of the 'white option' and the large-scale downsizing that was previewed in Chapter 1. Although touted as a solution to the overcapacity problem, and one that would stabilize employment at the company's other divisions, this has not been borne out. Instead, throughout the latter 1980s and into the 1990s, the lay-off situation at the Allan division has become progressively worse. Thus, for both 1993 and 1994, the company budgeted for only twenty-two weeks of production at the facility.[47] In effect, a regime of seasonal employment had been introduced at the mine, whereby workers would get some work in the autumn and spring, followed by lengthy lay-offs over the winter and summer months.[48] In our sample, over the years 1990–3, employees had been out of work for an average of 38.9 weeks. Almost 30 per cent of the sample, though, had experienced 60 or more weeks of lay-off over this three-year period.

To this situation other irritants have been added. Downsizing at the Cory facility was so complete that the only way production schedules could be maintained was through the performance of large amounts of overtime work. Incredibly, everyone in our sample at Cory had been asked to work overtime in the year preceding the interviews. This translated into a median figure of fifty overtime assignments per respondent in the 1994–5 production year, an amount of overtime that is far in excess of what has been reported at the other potash mines. This extra work has perpetuated the ongoing maldistribution of employment at the company as a whole, with predictably divisive effects on the union.

Additionally, and in a reversal of the policies that prevailed prior to 1982, the company has made much more extensive use of outside contractors for mine development work and non-routine maintenance jobs.[49] Typically such contracting out occurs just before lay-offs take effect, but after notices of shut down have been served on the workforce. The effects of this practice on workforce morale can easily be surmised.

As a result, the introduction of a 'turnkey operation' that is attuned only to the global supply and demand curve has posed some problems for managing the workplace, not to mention the now subemployed workers.[50] Declines in productivity and more frequent equipment breakdowns following the notification of lay-off were brought to my attention by mine management. So too was a general lethargy associated with the start-ups that followed the lengthy, indefinite lay-offs.[51] The latter is partially attributed by management to lax supervisory practices, especially after long shutdowns, when front-line staff may exhibit some sympathy for returning workers. This and other practices have led to what management at PCS, in an unconscious appropriation of Marx, refer to as 'theft of time,' work time that employees 'illegitimately' snatch from the company![52]

Potentially more serious is the retention of skilled trades personnel over the course of lengthy biannual lay-offs. To remedy the problem of turnover associated with a market-driven production regime, and following a worker demonstration over the lay-off issue at company headquarters, PCS introduced a return-to-work bonus program in 1991. Under this measure, workers, *who return* to the company's employ following each indefinite lay-off, receive $200 for each week of lay-off. The money is received only *after* recall. This is also a discretionary form of aid, as the company refuses to place the provisions of the program in its collective agreement with the union.

According to management sources the return-to-work bonus has stabilized the labour force in the wake of a transition to radically marketized relations.[53] For the union, on the other hand, the whole situation has been paralyzing. As poignantly observed by one member, the return-to-work bonus 'negates lay-offs as a strike issue.'[54] Instead, the whole matter of a transition to seasonal labour has been individualized, and in some cases, experienced as the loss of homes and other possessions. Motions to ban overtime, or to impose penalties on members who accept it during lay-offs have been judged *ultra vires* the union's constitution.[55] This has done nothing for solidarity either within the local or between this local and other PCS bargaining units, which may go on producing full out while Allan workers endure unemployment.

Is it then possible to refer to PCS as a market modified or 'late' variant of a Fordist employer? This appears to be the most reasonable designation for the following reasons. First, the hallmarks of Fordism, including job control unionism and division of labour, remain as important as ever at PCS. Downsizing activity and the utilization of seasonal employment have been filtered through the structures of seniority, CWS, and a well-exercised grievance system. Indeed, one aspect of radically marketized industrial relations not yet mentioned has been a general crackdown by the company on its version of the 'theft of time' problem. This has entailed a more rigid monitoring of lunch breaks and a frequent use of formal disciplinary measures to curb alleged minor infractions.[56] Such day-to-day practices are best located within the structures of managerial control that we have already reviewed.

Moreover, any pretence to employment security has been dropped, irrespective of the company's financial health. A Japanese-style commitment to job security for a core workforce is simply missing at this company. Instead, as with other Fordist employers, labour is accommodated through an enhanced internal welfare program that is tagged on to the minimal standards of the national unemployment insurance entitlements.[57] This does provide the company with a certain form of flexibility, mainly at the expense of labour. However, it would not be accurate on account of this to place PCS on the post-Fordist end of industrial relations. According to the flexibility theorists, numerical flexibility is reserved for a peripheral labour force, while job security and functional flexibility are offered to permanent core workers.[58] This has certainly not become the lot of employees at PCS. Here, a core labour force has lost ground, in typically 'late' Fordist fashion. Planned unemployment is used as a strategic lever by the company in the production politics of the workplace.

The designation of PCS as essentially Fordist can be used as a theoretical benchmark, against which other practices may be analysed. This is done by invoking a qualitative index of dissimilarity, beginning with firms that are more similar to PCS and then moving on in the direction of employers that are increasingly dissimilar, that is, more post-Fordist in nature. This will permit the systematic controlled comparisons of later chapters. Next we turn to a case which is quite similar to PCS, apart from one very important exception.

Cameco and Radically Spatialized Industrial Relations

Cameco is a relatively new entity, spun out of the former federal and provincial crown corporations of Eldorado Nuclear and the Saskatchewan Mining and Development Corporation in 1988 as part of a privatization bid. By virtue of this amalgamation, the new company immediately attained the status of world's largest uranium producer, with about 16 per cent of global production, or 55 per cent of Canadian production at the end of 1993. The first share offering to private owners was made in 1991, when ten million shares were sold off in a few days. At the time of writing, fully 90 per cent of the company was in private hands, while a residual 10 per cent was still held by the government of Saskatchewan. Cameco shares are traded on the Toronto and Montreal exchanges. They have advanced from the initial sell off price of $12.50 in 1991 to over $31.00 per unit at the start of 1995.[59]

Currently the company is operator and majority owner of two active uranium mines, Key Lake and Rabbit Lake, as well as the Blind River and Port Hope conversion facilities.[60] Cameco also holds majority interest in two other mines which are both at various stages in the environmental impact assessment process, McArthur River and Cigar Lake. The McArthur River project is the one that is furthest along in development. Providing it does come on stream, and at the moment this seems assured, it will be the world's richest uranium mine. In addition to these holdings, Cameco is also a one-third owner of Crow Butte Resources Ltd., which operates a uranium mine in Nebraska.

The company has shown a propensity to diversify into other metals in recent years. Gold production commenced at Contact Lake, Saskatchewan, in 1994, a development in which Cameco holds both a two-thirds interest and operator status. Also in 1994, the company became the chief minority owner and operator of the Kumtor gold mine in the Republic of Kyrgyzstan of the former Soviet Union. The latter is one of the world's largest known deposits of gold.[61]

This study covers one division of Cameco's operation, the Key Lake mine and mill complex. Production on site commenced in 1983 in an open-pit operation which remains the world's largest uranium mine currently in operation. At the time of writing, Key Lake directly employed 399 workers, of which 285 were in operations or maintenance jobs. Although mining of the ore body will come to an end in 1996–7, the mill at Key Lake will continue to provide service for the new McArthur River operation after 1999.

Owing to the fact that Cameco, in its present configuration, is of recent origin and that it was formed through a merger of existing companies, a common industrial relations approach has just begun to gel. According to management, the priority to date has been placed upon smoothing out policies between the company's different sites to create some internal consistency.[62] Although the company has recently entertained some discussion apropos moving towards Total Quality Management and gainsharing frameworks, and has heard pitches from different consulting firms in relation to such programs, no further action along these lines has been pursued. Certainly, given the company's performance since privatization, there has been no sense of urgency around the adoption of new production relations. Rather, as the manager of the Key Lake operation noted, firms that have opted for new industrial relations have generally done so in response to some experience of trouble. At Cameco, management has not perceived this dynamic to be of much relevance.[63]

Given this definition of the situation, Cameco also falls towards the traditional side of labour/management relations but with some interesting twists. The absence of an overarching program that would readily frame production politics at Key Lake is itself telling. In other words, there is no equivalent to the Loss Control program found at PCS, that is, an anchoring strategy around which managerial efforts are directed. Instead, location and the management of spatial relationships provide this attachment. The importance of transportation, of getting the workforce to the mine, and the management of civil society, which is literally one step removed from the labour process, is what lends the employment relationship its real specificity at this site.

These feats are achieved through the prerogatives of managerial authority that may be used in both paternalistic and in less benign ways. Thus, although many of the features of employment at the mine are comparable with the other sites, they are often overlaid with a degree of managerial control that is unique in a unionized mine setting.

The job classification system that structures the employment relationship provides a first example. In total there are thirteen job classes at Key Lake. As at PCS, these categories determine tasks, rates of pay, and progression. In some cases they are even more detailed than at PCS. For instance, there are five equipment operator positions in the pit, each one associated with a specific job class number. Occupation of a specific position involves the use of certain pieces of mobile equipment, but not others. An equipment operator two, for example, may perform functions on a crane of under twenty tons, but not on larger cranes. An equipment operator five may drive some of the vehicles in the field (buses, fuel trucks, front-end loaders), but not others (haul trucks, water wagons, garbage trucks, and so on). Notably, however, this gradational classification system has been arrived at in the absence of CWS. As a result, the union does not play a co-determining role in the evaluations of jobs or premiums. Pay scales and job class differentials are, of course, negotiated, but there are no formal rules of job evaluation to which the union can appeal. It is the company that *singularly* stakes out the comparative worth of each task.[64]

A diminished union presence is detectable at Key Lake in other, often subtle ways.[65] As noted earlier, owing to the remoteness of the mine site, management controls access to the property and other forms of communication in a much more visible way. Although it is not a technological necessity, the local union has only received access to its own phone or fax lines to the site as of the most recent contract. Prior to this, phone conversations with local union officials were often conducted within earshot of supervisors. As well, in the past, access to the site has been made cumbersome for union officials such as the staff representative, who has either had to drive several hundred kilometres along unpaved northern roads, or pay for a seat on the company jet. Finally, and as is common practice at the other sites in the study, the local union and its bargaining committee do not receive paid bargaining time from the company. Although this may not be a seminal matter, it is indicative of a specific climatic that is lent to production politics at the mine site.

Gleanings of the managerial ethos operant at this site are available in other places as well. While we were conducting our interviews, management unveiled a new Absentee Corrective Action Program. This instruction outlined the 'actions to be taken when excessive absenteeism occurs to ensure the root cause is determined and corrective action is taken.'[66] This entails a detailed process of reporting and monitoring, whereby workers are singled out for *non-voluntary* 'treatment' if their absenteeism rate is above the site average for any quarter, or year, or if the employee has two or more absences per quarter. Thus, even workers with reasonable work attendance records may find themselves in the program, for the simple reason that they have had more absences than their peers. In such instances, the worker is enrolled in the program by the company and is expected to develop an 'action plan for improvement' with his or her supervisor. Further absences are reviewed with higher levels of management, right up to the mine superintendent, and may eventuate in an 'employment suitability review' being conducted.

More generally, this approach may be taken as part of the company's overall philosophy of discipline where: 'Positive discipline is a vehicle designed to hold employees accountable for their actions while providing them an opportunity to adjust to the corporation's policies and requirements. Positive discipline is not designed to punish ... but rather to correct and redirect employee behaviour toward action in line with corporation standards and regulations.'[67]

Such statements have a definite coercive tone to them. By positive discipline, the company has in mind a calibrated set of correctives that are intended to bring employee behaviour into line with managerial objectives. Incidents are thereby classified in accordance with their perceived seriousness, ranging in nature from minor offences, through to the fifth level of intolerable actions leading to discharge. For each gradation a 'step' is meted out to the offender; the steps are both cumulative and subject to a statute of limitations. That is, if a worker is already on step one, perhaps for being late for work, and then receives another step, the employee goes to step two. With a clean record for three months, the hypothetical employee moves back down to step one and, barring further incidents, to a clean record in another three months. More serious offences such as uttering verbal threats, refusing to do assigned work, or engaging in unsafe behaviour, to take three examples, evoke higher levels of discipline, which take longer to remove from the worker's record.

The point is not that Cameco stands alone in using such systems. This is not the case. Rather, management's preoccupation with such issues as absenteeism and discipline may in large degree be related to the central feature of the whole Key Lake operation, namely, the levels of spatial coordination that are required to organize modern forms of production in a remote environment. The fly-in/fly-out system of workforce organization is a novel way of conducting production. Although not uncommon for neoteric managers, this, to our knowledge, was one of the first examples where whole workforces were picked up at various locales, across a large catchment area and shuttled across vast distances for relatively short periods of time, in order to perform their work. Cameco was the first corporation to use this system as an alternative to permanent settlement (company towns) adjacent to the production site. Since 1975, such radically spatialized systems have become the industry norm.

Not having the supporting structures of civil society immediately available, the organization of time and space attains a significance for management that is not likely to be found elsewhere.[68] To begin with, workers must literally be brought together at the point of production. Unlike the other mines, this is a managerial function. In this case, company-owned aircraft, which are contracted out to a charter airline, pick workers up four days a week from various points across the south and north of the province. Flying time, including weather and other delays, is not paid for. Once in, workers report immediately for work, beginning a seven-day, twelve-hour work roster, prior to flying out again for a seven-day respite.

On site, about three kilometres from the mine and milling activity, the company has constructed a 300-unit residential facility, which includes single rooms for the employees, a cafeteria, and such recreation facilities as a gym, racquetball court, sauna, and tennis court. Early on in the history of the 'settlement,' and through volunteer employee labour, a curling rink had also been constructed. Most workers on site share rooms with their opposite numbers on the alternate week's shift, which inhibits individual customizing of rooms, although in a few instances, where a room has only a single occupant, it may reflect the resident's interests or hobbies. Women workers inhabit their own wing of the complex, and it also seems as though management has been assigned to certain floors, although the rooms remain uniform throughout the complex. Operations such as housekeeping and catering have been contracted out to local (that is, northern) business ventures and these workers also live in the same complex.

Managing the camp and transportation to it is an important aspect of mine operations. A missed plane, or late night rowdiness in the camp can have immediate effects on company production schedules. For these reasons, the company has built in certain safeguards to the management function. For example downsizing has not been pursued to the full extent that technological development would permit. A case in point is to be found in the mill. When mill functions were centralized in one control room, redundant workers were allocated to a newly formed 'day crew,' which in addition to providing housekeeping functions for the mill, also formed the basis for a workforce reserve in the event that other workers failed to show up for their scheduled week in.[69] Consciously, and in a typical Fordist fashion, a workforce buffer was built into managerial calculations.

Space can not only be conceptualized as a barrier to be overcome, or annihilated by capital, as Marx put it, but also a resource that is available to management. Let me provide some examples taken from the site. In 1990 workers engaged in a short wildcat strike in protest over the slowness of contract negotiations. The company retorted by threatening to evacuate the property and made it known that the plane ride south 'would be the last one that the workers took.'[70] Although this response was not the sole reason for bringing the action to an end, there is no doubt that it did have a 'chilling' effect on the situation. The previously mentioned union reliance on company fax and phone lines, as well as on permission to bring union representatives on site, is another instance of localized power that emanates from control over space and distance.

The unique spatial relationships that envelop industrial relations at this site also have implications for the reach of discipline and company power. Because workers live collectively in company auspices, labour/management relations extend considerably beyond the proverbial 'factory gate.' Thus, 'As discipline for both work related and *off-duty* activities are integrated, disciplinary steps are cumulative irrespective of the nature of the discipline [my emphasis].'[71] Offences that are covered in the company's document on the administration of discipline include such items as the wearing of dirty work clothes in the residence, the possession of liquor outside of the social club (that is, in individual rooms), and noisy conduct after 11:00 PM. As one might expect, a fair amount of conflict in the form of disciplinary/grievance action revolves around life outside of work.

Yet another poignant reminder of this extension of company influence on the lives of employees is contained in the periodic missives workers received at their home addresses, advising them of the undesirability of taking on second jobs. For some workers, this is a major advantage of the week in/week out system of production. It can often be combined with a second job, which is carried on in the home community. From management's perspective, however, second jobs can interfere with company needs and requirements, as when workers incur injury while at another job, or show up at the mine site tired and unprepared for work. For these reasons, Cameco actively discourages the holding of a second job, even as it touts the flexibility that its production schedule offers workers. This is another example of the extension of control that inheres to radically spatialized work relations.

A third and final consideration, that radically spatialized industrial relations impinge upon, is the existence of the union itself. In many ways there is not one workforce at Key Lake, but many. In addition to the dual market that is generated by race and skill factors, the simple fact is that there are two other workforces, namely, week one and week two, which have only the slimmest connections with one another. Half of the workforce is always absent, dispersed in numerous communities within and out of province, while the other half is on site. These workforces 'meet' only once per week for short crossover periods. For those who are at work at any given time, there is the additional subdivision of day and night shifts. Thus spatially and temporally there are four workforces at Key Lake. This is reflected in the necessity of convening four meetings in order to constitute one inclusive meeting of the local union!

By selectively drawing upon numerous communities to construct a labour force, Cameco and the fly-in/fly-out system of work organization has succeeded in producing a spatially and temporally contrived 'community' at Key Lake. This poses greater-than-average challenges for the construction of union identities that are overlaid by issues of race and residence. The social designators of 'northern,' 'southern,' Aboriginal, and non-Aboriginal assume a critical importance at the site of production. This is reflected in local union divisions which are unique and that reverberate in the style of management found at the worksite. More specifically, if the union can be portrayed as a 'southern,' 'white' institution, principally interested in defending the skilled trades, the company can use such perceptions to its own advantage. This becomes especially pertinent as operations are being scaled back with the exhaustion of the Key Lake ore body, and the question of seniority looms larger for the future. Meanwhile, through an active educational program and some high-profile defense cases, the union has attempted to overcome some of these obstacles to internal unity.

In recognizing the uniqueness of radically spatialized relations for work organization, I do not want to anticipate outcomes or overplay potentially negative features. It is fair to say that some workers value the existing shift system for the blocks of free time that it provides them, especially in combination with holiday time. Others find camp life to be a liberation from unwanted domestic responsibilities. The company has also committed resources to sustain what it considers to be a reasonable quality of life in such a context. Nevertheless, management personnel and co-workers alike frequently admit that such a regime is not for everyone. Some, for example, are clearly resentful of the lost family time (missed birthdays, weddings, et cetera) that has accumulated over the years. For present purposes, it is sufficient to highlight the determining effect that the spatial domain exercises on operations at this site. This factor contributes to and reinforces a commitment to traditional Fordist managerial practices. Given this synopsis, reflections of what Burawoy has described as a migrant labour system, a 'company state,' or employer paternalism might be ascribed to Cameco.[72] Describing an earlier time, Burawoy states: 'Through their control of housing, provisions, company stores, education and religion, masters were able to consolidate their rule in all spheres of life ... The company state went beyond market despotism to intervene coercively in the reproduction of labour power, binding community to factory through non-market as well as market ties.'[73] Intriguingly, this theme comes up again in Burawoy's treatment of the copper mining industry in post–Second World War Zambia. In this example, the company in question constructed various recreational venues for the resident workforce, including beer halls, dancing societies, and other clubs as part of a social technology of total surveillance.[74]

Under certain conditions, according to Burawoy, a company-state factory regime could also lead to corporate paternalism. In such instances, the authority of the owning class 'permeated not only public life, but also the day-to-day existence of their hands beyond as well as within the factory.'[75] Through their beneficence, 'employees came to identify with the fortunes and interests of their employer.'[76] Unlike the company-state formation, which does not brook any form of independent working-class organization, corporate paternalism is capable of carving out a subordinate place for trade union existence.

There are residues of these features, as well as some superficial resemblances to the migrant-labour system phenomenon, that Burawoy and others have discussed at Key Lake. The extension of managerial codes of discipline to life in the camp is suggestive of Burawoy's company state. Management's role in organizing blocks of civil society at the property, such as in co-sponsorship with the union of various clubs and recreational activities also invites forms of paternalism. Additionally and unmistakably, the uranium industry is a political 'hot potato.' Company officials frequently approach the local union for aid in presenting 'the industry's case' to governments and assessment boards. The labour movement's close ties to social democratic governments is viewed by the industry as particularly propitious when industry lobbying efforts need to be organized. Finally, work at Key Lake is characterized by a radical separation of daily wage-earning activity, or what Burawoy terms maintenance functions of labour power reproduction, and working-class household renewal.

Despite these parallels, industrial relations at Key Lake fall under neither the company state, the corporate paternalist, nor the migrant labour system paradigms. The company state regime is thoroughly unilateralist, and although the local union at Key Lake may be weaker than its counterparts in the potash industry, it is nevertheless a countervailing power at the site. Furthermore, precisely what is missing at Key Lake is the type of community that would allow company-state power or paternalism to fully take hold. Unlike real company towns, or even the family compounds that characterized sections of the African mining industry, the radically spatialized work environment has no true community appended to it. Without such, the powers described by Burawoy in the company-state and corporate paternalist regimes cannot be completely operationalized. Thus, even if the parallels are interesting, it would clearly be a misnomer to consider Key Lake a modern-day counterpart to the company-state or the corporate paternalist models.

The same holds true for drawing analogies with migrant labour systems. A key component of this form of work relation is the *forced* migration and repatriation of workers between spheres of household activity and wage employment. This is occasioned through the selective use of state power, principally in the denial of citizenship rights for migrant workers. Although there is no denying that economic circumstances drive many northern residents to seek out work in the uranium industry, direct political coercion, *force majeure*, does not enter the equation, as it has in the cases of historical migrant labour systems. For this reason I have used the notion of radically spatialized industrial relations to capture the chief qualities of work in the modern uranium industry, as well as to distinguish this case from earlier regimes of industrial relations that were associated with direct colonialism and mining.[77]

One should also not infer that there is no experimentation at Key Lake in alternative human resource management. Several instances of such were uncovered during our investigations. When the company was privatized, shares were set aside as part of an employee stock purchase plan, wherein the company matched dollar for dollar employee investments up to pre-set limits, which were determined by the individual's wage. This was a popular move, although some workers did not approve of tying the company's contribution to each individual's current wage, as this privileged higher income earning staff. In addition to the one-time share offering for workers, there is a maintenance problem-solving team on site composed of one representative from each of the trades, an employee from the warehouse, and the general foreman. Established at the instigation of the general maintenance foreman, this group convenes three or four times a year to discuss weak areas of the operation as well as opportunities for improvement.[78]

Local management is also pressing for more workforce flexibility in mill operations. Presently, there are five job-class levels in the mill. In proposals that went before the union bargaining committee, the company asked for a revision of this structure. Under the proposed reform there would be a five-year training period that all mill employees would be subject to, after which workers would be given the top wage rate for mill operatives.[79] This would be set at one increment above the existing wage ceiling in the mill. In return, the company would expect greater flexibility in its deployment of workers around this department. This, of course, is recognizable as a so-called multiskilling/payment for knowledge initiative.

These are not insignificant forays. Nevertheless they remain both limited and piecemeal. Thus, multiskilling proposals are intended to cover just one department, mill operations. Other groups who have expressed an interest in more advanced training, or in greater prospects for internal departmental mobility are not included in the company initiative on multiskilling.[80]

The one problem-solving team that exists on site is limited in its membership, largely because management does not want to spare workers from their normal work routines.[81] This priority has ruled out any forms of team training, or educational initiatives into team problem-solving methods. Instead, training exercises remain quite conventional and similar to the program in effect at PCS. Individuals are pulled off of the job when it is opportune, in order to receive self-instructed modular training on a CD-ROM type of system. Units covered for mill operators include such topics as basic mathematics, principles of measurement, diagrams and graphing, pump technology, and the techniques associated with the sampling of product when it is in midstream. Each field is broken down into various units that workers are expected to complete. One interesting side dimension to this is that now that training has been individualized and computerized, company trainers possess detailed records on the performance of each trainee. Time spent learning each module, and completing examinations and individual questions at the end of a unit, is automatically calculated and stored in the training program.[82]

Finally, although workers were included, in a very limited way, in the stock sell off that accompanied the privatization of Cameco, this was a one-off initiative. Currently no plans are underway to develop gainsharing or other programs that would directly link earnings to work effort.

In each of the foregoing domains of organization – multiskilling, utilization of team approaches and training in this paradigm, and alternative compensation programs – change has been distinctly limited and at best piecemeal. Such circumscribed, ad hoc, departures from Fordist norms have not been part of a larger integrative strategy that would tie such measures together into a post-Fordist package of industrial relations reform. To date this has been precluded by virtue of the prosperous state of the industry and the opportunities for control that are afforded management through the unique spatial characteristics of the operation. It is precisely managerial control over the organization of spatial relationships which becomes the dominant theme to emerge at Key Lake. What impresses is the extent of this control.

The next chapter examines two companies that have felt compelled to distance themselves to a greater degree from the Fordist principles that characterize the organization of work and production politics at PCS and Cameco.

4

Corporate Cultures of Employment II: Two Post-Fordist Firms

The two remaining companies differ from their corporate cousins in so far as they have made moves away from the Fordist protocols that define the organization of work at PCS and Key Lake. Of course, what I am referring to are matters of degree. Thus, the notions of Fordist and post-Fordist industrial relations will continue to be used as heuristic markers for social trends and not as reification. As social trends in management, there is nothing sacrosanct about them. What is in vogue today (yesterday) may be already passing into disfavour tomorrow (today). So, with these qualifications in mind, we can continue. Central Canada Potash Ltd. and Agrium, the two producers examined in this chapter, deviate in varying, but important, ways from the firms inspected in Chapter 3. That is, in certain respects they have begun to diverge from the Fordist paradigm of organization. As to whether they will ever arrive at something called post-Fordism, like elsewhere, the verdict is still out.

Central Canada Potash: Elements of a Passive Post-Fordism

Until recently Central Canada Potash (CCP) had been a wholly owned subsidiary of the natural resource giant, Noranda Inc. Historically, Noranda has been a major player in the base metals field, in forestry, and in oil and natural gas. Its name is associated with ownership, or major stakeholder status in such well-known companies as Falconbridge, Kerr Addison, Hemlo Gold, Canadian Hunter, Norcen Energy, and until 1993, MacMillan Bloedel.[1] In short, Noranda is the quintessential Canadian multinational resource company.

In what may be symptomatic of our times, over the course of this study ownership title to Central Canada Potash changed twice in less than a year (1994–5). First, the property was sold by Noranda to the much smaller manufacturer, Vigoro Corp, an American-based fertilizer producer headquartered in Chicago.[2]

Within a matter of months Vigoro went on to merge with International Metals and Chemicals Ltd. (IMC Global), the second-largest potash producer in Canada. The merger will double IMC's property holdings in Canada, giving it a total of four mines in Saskatchewan and adding substantially to the 19 per cent of industry capacity that the company already holds.[3] In addition, IMC is America's largest producer of concentrated phosphates, with ten mines currently held through IMC-Agrico in the United States.[4]

Rumours of Noranda's sale of Central Canada to Vigoro were widespread at the time of the employee survey at CCP. When the mine was first constructed in 1969, the minority owner was an American farm cooperative. This arrangement provided CCP/Noranda with an assured market for its potash in the American midwestern farm belt. In 1978, however, the partnership was terminated and Noranda found itself the sole owner of CCP, without any sort of specialized marketing division.[5] It is within this context that the subsequent ownership changes have occurred, whereby Noranda unloaded the one potash mine from its corporate portfolio. Notwithstanding these changes, employees were unsure of what to expect. They could only respond to our inquiries as Noranda employees, which is what they still were at the time of the interviews. As shall be seen, the changes in ownership are still having consequential effects on employment relations at the mine. So far this has mainly involved the winding down of initiatives that Noranda management had sponsored throughout the 1980s, which is where the analysis commences.

A number of respondents identified a two-day wildcat strike in 1981 as a watershed event in the history of relations at the mine. The strike, triggered by the dismissal of a worker and shop steward, would lead to prolonged litigation. It also served as the catalyst for a number of uncoordinated attempts to move beyond the highly authoritarian workplace relations that characterized employment at CCP.[6] The first tangible outcomes of this episode were the replacement of several 'old line' managerial cadre on site, and a movement towards the initiation of an *'Employment Involvement'* program. The latter was sold to CCP as a project 'specifically designed to improve the willingness to work, understand, and accept.' Through employee involvement, management could 'make better use of the skills and knowledge for which we already pay.'[7] Initiated by means of a general workforce survey that revealed both physical and sociological problems within the operation, this exercise led to the brief establishment of a Human Resources Steering Committee late in 1984.[8]

Most important, from our perspective, were the employee proposals that emerged from this first climate survey. They included the introduction of profit-sharing and stock purchase plans, better training, and improved employee performance reviews.[9] These items would later come to fill out the company's

TABLE 4.1
Management rights under proposed CCP Employee Involvement Plan

	Negotiable	Non-negotiable
Capital expenditures		X
Approval of expenditures	X	
Results from each employee		X
Respect from fellow employees		X
Safe working conditions		X
Cooperation among employees	X	
Hiring	X	
Standards for education	X	
Feedback	X	
Setting goals/priorities		X
Day's work/pay	X	
Major organizational decisions		X
Decisions on work methods	X	
Veto power		X
Hours/schedules		X
Production standards		X
Performance standards	X	
Corrective action		X

Source: Central Canada Potash, Human Relations Steering Committee, 15 May 1985.

reform agenda. The Human Resources Steering Committee (HRSC), on the other hand, was short-lived. Chartered to promote employee involvement through the establishment of problem-solving teams, in practice, the committee was restricted to giving effect to the mine manager's statement on human resources issued as a sitewide press release early in 1984.[10]

The concept of employee involvement as developed by the company was limited in both theory and practice. At one point the HRSC undertook to scope both management rights and employee rights in the new workplace. The results of this exercise are displayed in Table 4.1, which is reproduced exactly as the committee finalized it. Although this reconfiguration of workplace relations envisaged some inroads into the realm of co-determination, with negotiable joint responsibility between management and workers for hiring, training assessment, and work methods, overall managerial authority was to be retained through control of capital expenditures and retention of final veto power over all proposals.

In practice, membership on the HRSC was stacked in favour of management, who initially had eight members to the union's two.[11] Any alteration in the composition of the HRSC required the general manager's approval, while in turn, the steering committee controlled the membership of any project teams that would come out of the program. From these parameters, a sense of tight managerial control over the change process is fairly obvious.[12]

Reluctant to become a participant in the first place, and citing the existing inequalities in the composition of the Human Resources Steering Committee, union involvement came to an end with a formal letter of withdrawal, a little more than a year after the Employee Involvement program was first launched.[13] The impression had been left that while CCP may have wished for greater *employee involvement*, this did not encompass greater *union involvement* in the affairs of the organization.[14] And without at least tacit union support, the Employee Involvement initiative quickly withered on the vine.

As the next foray into alternative industrial relations at CCP was to suggest, the earlier Employee Involvement scheme had a number of factors working against it. Aside from the platitudinous statement by the then mine manager on 'human relations,' the company had no clear notion of what it was trying to accomplish by moving into a 'jointness' model.[15] More to the point, frequent lay-offs, combined with ongoing technological change were hardly propitious conditions under which to elicit a heightened sense of employee cooperation. Beginning in 1982, the lay-off situation was to worsen until 1987, with the last lengthy shutdown occurring in 1988. These lay-offs were similar in nature to those experienced by PCS workers; one or two per year, each of several months duration. Partly fortuitously, and partly owing to the changes in PCS production and marketing strategies that occurred under radically marketized production relations, lay-offs at CCP have since become avoidable events.

By the end of the 1980s there was also a pause in the latest wave of technologically induced downsizing. Changes in cutting practices at the mine face, pursuant to the introduction of superior face bit patterns (full face cutting patterns), and fully automated and enlarged skipping capacity entailed job shedding underground. Shift numbers in underground production shrank from complements of twenty-four workers in the early 1980s down to totals of thirteen workers, while the centralization of automation in the mill cut in half (thirteen to seven), the number of operator positions required to staff this area.[16] Meanwhile, across the site, approximately thirty trades positions were lost over the same period.[17] The greatest redundancies to date were thereby recorded and were completed by the end of the 1980s. This abatement in downsizing, along with the cessation of lay-off activity, provided the company with another opening at the end of the decade for new employee participation schemes.

Coinciding with the end of lay-offs in 1988 was a rapid announcement of new initiatives at the mine. These included the undertakings as listed in Table 4.2, and

TABLE 4.2
Chronology of reforms at Central Canada Potash, late 1980s to early 1990s

Year	Program
1987	Employee Recognition Program: recognition of individuals and groups for contributions to company's success.
1988	Performance-Sharing Program: payouts based on profits, productivity, and safety record.
1988	Employee Stock Purchase Program: subsidized purchases of CCP stock up to pre-set limit.
1989	Company Vision and Values Statement: mission statement.
1989	Renewal of Climate Surveys: outside consultant.
1990	Parallel Apprenticeship Program: in-house apprenticeships.
Circa 1992	Institution of Peer Review Groups, Quality Teams & Employee Action Response System: measures in support of team approach.
1992	Voluntary Severance Package: two weeks' pay for each year of service and up to two years' paid tuition for new training.
1993	Innovation: joint management/union team.
1993	Education/Training Windows: joint program with Employment Canada: educational leaves funded by unemployment insurance and company top-up.
1994	Training Needs Assessment Project: cross-training initiative.

discussed here roughly in the order of their chronological appearance. In reviewing these measures two observations are relevant. First, several of the initiatives were either introduced from off-site or were in keeping with practices that had been adopted by Noranda on other properties. In some instances, then, CCP management was bringing policy at the mine into conformity with parent company objectives.

Second, and more puzzling, is the apparent absence of any logical chronology to the unfolding of events. For instance, it might have seemed more reasonable to conduct the second round of climate surveys prior to issuing a mission statement for the mine and its workforce. But this was not done. Instead new initiatives, such as the performance-sharing plan and the penning of a mission statement, preceded the information gathering stage of the climate surveys. Similarly, it would have made more sense to have introduced new reforms at the company in such a manner, so as to at least be seen to be lending substance to the vision and values charter, but again this defining statement came after some of the more substantial reforms.

Be that as it may, that the history of change at this operation was not so neat may in itself be suggestive. It could, and probably does, signify that management remained in an experimental mode, in the strict sense of the term. Recognizing the importance of employee attitude, what was being sought was a better motivated, highly committed workforce, and to that end various strategies were being 'tried out.' Over and above this, there was no long-range master plan. Equally important, the ad hoc nature of their chronology underscores the point that these were *management* programs, again, in the strictest sense of the term. The reform agenda was being run from the top down, although it was hoped that individual employees and the union would 'buy' into the proposed changes.

Credence to these remarks is lent by a more detailed look at one of the first and arguably the most important innovation, the performance-sharing plan. The introduction of profit sharing at Central Canada was tied to the search for better workforce morale on site, but more concrete goals or targets would only be defined at a later stage of the change process.[18] Thus, there was an intuition that by giving employees 'stakeholder' status through a share of divisional profits, performance might improve, although by how much or how little, no one was prepared to predict.

The introduction of performance sharing was also an uneven affair. Because it includes all employees, unionized and out-of-scope alike, the plan is not part of the collective agreement. Nevertheless, union cooperation was sought to attain the desired objectives and, as a result, union representation was allotted for the committee to oversee the detailed planning and yearly operation of the program. In early drafts, however, it was proposed that the performance-sharing committee, composed of three union and three staff members, would be elected by all plan members. From the local's perspective this constituted a grave difficulty, giving management a vote on union committee representatives. Only after a threatened unfair labour practice charge was leveled against the company and *its* plan, was this feature altered to allow for sole union appointment of its representatives and staff designation of its agents.[19]

Participation in alternative remuneration schemes is definitely one of the controversial elements in the post-Fordist agenda. This was reflected in the cautious advice and lukewarm reaction that the local union received from the national office on the CCP performance-sharing plan. Two items of concern are particularly worth noting. Under the terms of the program, either party to it could back out after providing one year's notice of intention to do so. Union officials warned of the possibilities that this extended to management for engaging in benchmarking activities. That is, after appropriating employee ideas and improvements, there was nothing to prevent management from unilaterally cancelling the program. As long as profit sharing remained outside of the collective agreement this could be one future scenario.[20] Additional job haemorrhaging was another concern. As the national office underlined, 'We can hardly agree to a plan that will make one member work harder so the company can lay-off another member. Lay-offs due to productivity gains must be prohibited.'[21]

Nevertheless, the performance-sharing plan has turned out to be more popular with the workforce than originally foreseen. Based upon a combination of divisional profitability (8 per cent of net profits), performance objectives that include increased tonnage over a 1970–86 base average, and safety records (injury rates as measured against industry averages), the plan has paid out increasingly large, indeed some would say unexpected, bonuses. The 1995 estimate, for example, was in the range of $7,500 payments for each employee in the company's hire.[22]

Introduced around the same time as performance sharing, and in the same genre, was the employee stock program. This was also a Noranda-wide initiative. Under its auspices, workers could set aside 5 per cent of their wages for share purchases, with the company matching an additional $.30 on every dollar so targeted. Workers could also devote an additional 5 per cent of their wages to company stock purchases, although this amount would go unmatched by the corporation. Although the intent was similar to performance sharing, namely to foster a greater 'stakeholder' status on the part of the workforce, the indications are that this program has not had anywhere near the impact, material or otherwise, that profit sharing has had.

The other program that had some interesting potential from the union's perspective was the apprenticeship training program that was announced in 1990. A common complaint amongst union locals in the industry is so-called hiring off the street. That is, companies will ignore any type of training that would significantly upgrade existent skill levels among those already on the company payroll. Instead, it is cheaper and less risky to simply externalize the costs of long-term training by hiring the skills on an as-needed basis from the external labour market. This, of course, denigrates the internal labour market available to the workforce. Rather than upgrading the training of workers who are already in situ, from operator to trades positions, management will hire from the external labour market ('off the street'), bypassing its own workforce. Opportunities for internal occupational mobility are thereby curtailed.

The apprenticeship training program was introduced to remedy this complaint. Under its provisions, CCP would offer one new apprenticeship per quarter to one of its own employees, with selection being based upon a combination of seniority and prequalifying competencies. A joint union/management committee administered the program, although the company alone determined the trades mix that would be offered through it. Successful applicants would then be seconded to an apprenticeship program where they would take the four years of required training and education. At the end of this period, if a suitable position existed, the graduates would be available for it, otherwise they would return to their previous position until something more appropriate became vacant.[23] For CCP, the parallel apprenticeship program, as it was called, offered another venue for future downsizing activities. In this regard, it would be easier for the more highly skilled workers to find alternative job placements than for the less-skilled workers. For the union, the program provided an opportunity to bring more work in-house. With skills on site, there would be even less reason to contract work out.

The direction that CCP was wishing to pursue is at least partially reflected in the statement of visions and values that the company produced in 1988. Among other things, it holds: 'People are our greatest asset and as such CCP will consider the human element in all its decisions. CCP will have the most qualified, motivated and committed people, recognized by our competitors as the best in the business.'[24] The same document references the importance of teamwork, stating, 'We believe that the best way to achieve our vision is through *Teamwork* at all levels of organization.' Innovation ('Our success depends on encouraging, championing and *Adopting innovative ways*'), training ('We will continually invest in the growth of our people through *Training* and development'), and product quality are other post-Fordist themes that are echoed throughout the text.[25] A commitment is also undertaken to provide quality work in a co-determinative environment: 'The vision can better be achieved with the commitment and *Active Participation of the USWA*. CCP will provide opportunities for *Work to be interesting, stimulating and rewarding*.'[26]

The scoping of a vision and values statement was again solely a managerial initiative, which it was hoped workers would commit to in time and through the support of such programs as the performance-sharing plan and the employee stock ownership scheme. Initially, the consulting firm that was brought in to help CCP develop the mission statement recommended the establishment of a joint steering committee with the union to oversee a new climate survey, as well as to act upon the results.[27] This suggestion did not curry favour with certain strata within management, who were neither enthusiastic about a shift in paradigm, nor with bringing the union in to the process.[28] In lieu of this, employee focus groups were arranged for each department with the objective being, as the consultants put it, to 'begin the process of getting employee involvement and buy-in to *their responsibility* in assisting CCP to meet *their* vision [my emphasis].'[29] Coming out of this, it was envisaged that each department would use the data obtained to develop specific goals and action plans, with the proviso that such objectives be congruent with the strategic goal-setting designs of senior management.[30]

If a label were to be attached to these recommendations and the process

TABLE 4.3
Model for participative management at Central Canada Potash

What is it?	Employee's experience	Leads to	Final results
Participation in goal-setting, problem-solving, and change projects	Increased control over work behaviour; Involvement in meaningful work	Increased acceptance and commitment; Increased security, challenges, job satisfaction, innovation and quality	Increased performance and productivity

Source: Adapted from 'Participative Management as an Ethical Imperative,' for Central Canada Potash. Reprinted by permission of the publisher from Organizational Dynamics Spring 1984 © 1984. American Management Association, New York. All rights reserved.

through which they were arrived at, Total Quality Management might be the closest approximation, although no formal references to this exist in the industrial relations audits that CCP commissioned. Unlike the previous Employee Involvement trial, the new initiative was, in the first instance, primarily focused on managerial practices and their reform.[31] Here, there was to be a renewed effort placed on managerial team building, dedicated to group problem-solving and collective goal-setting practices. Thus, the recommendations of the company's consultants were to 'first initiate team building ... with the management team. Later it may be feasible (and only on the recommendation of a specific manager) to do team building at the next level.'[32] Once again, strategic goal setting was to be solely a management affair. Plantwide priorities, as defined by senior management, would filter down to specific departments and units, at which time worker 'participation' would be solicited. Only at this stage, the consultants suggested, would it be appropriate to constitute a joint steering committee, composed of managerial and supervisory representatives, shop stewards and select, 'high performing employees.' In terms of a formal model, we may refer to Table 4.3, which was contained in the outside assessment that was done for the company. It purports to summarize the objectives and dynamics of the participatory model.

In order to activate what was now being referred to as a participatory management model, special emphasis was placed upon a catalogue of appropriate management skills, combined with supervisory training and socialization into the new management style. This, it was calculated, would be conducive to eliciting the greater employee initiative that management desired. The design was clear. Armed with an appealing vision and values statement, and the data provided through focus groups, management would elicit worker commitment to the realization of newly established goals. Higher levels of worker control over more meaningful work would eventuate in better corporate performance. The actual results, however, have been somewhat more pedestrian than the grand designs envisaged by the human relations model builders.

Following on the heels of the performance-sharing plan, other initiatives, which would lend a post-Fordist vision to potash production at CCP, included several endeavors to support a team-oriented environment. Peer review groups have been instituted around the introduction of new or special projects on site. The name chosen for these teams is somewhat misleading, as their function relates to the problem solving that is often associated with new technologies and methods, rather than with the evaluation of co-workers. At Agrium, discussed later in this chapter, these groups are referred to as special project teams, which is a more apt descriptor. At CCP, membership on such teams is composed of those trades persons, operators, and engineers who are directly involved in the use, design, and adaptation of new pieces of equipment, or special projects.[33] In the recent past, such teams have been used to adapt existing technologies to new ground conditions in mining, in the design of new underground shops, and for the installation of new dust collection technologies in the mill.[34] Recently, one such team was still functioning around a mine-borer automation project.

Quality teams would also appear to have the same casual, fluid structure as the peer review groups. A number of these groups were created in 1992, specifically to deal with perceived quality problems in the mill and loadout areas. Meetings of operators, supervisors, and engineers occurred on a regular basis until the immediate problems were resolved. These groups have become semi-dormant, convening only a few times a year.[35]

An additional adjunct to employee involvement also introduced during this phase of the reform process was a suggestion program. Known cutely as EARS, for Employee Action Response System, this measure, like similar designs elsewhere, has been intended to elicit cost-saving ideas from mine employees. A counterpart to this is to be found in the company Employee Recognition Program. Both individuals and groups could be nominated for recognition of contributions to the company's vision and values statement. Nominations are vetted by committee, using weighted criteria that include contributions to innovation, productivity, cost reductions, safety, morale, environmental stewardship, and community service.[36] Awards include the receipt of plaques, gold watches, and special recognition dinners.

A greater degree of formalization was brought to these activities by the creation of an innovation team in 1993. This seven-member body, comprising four union delegates and three management representatives, was charged with identifying 'possible roadblocks to idea generation and idea acceptance' at the site. Given this mandate, the innovation team sought to 'seek input from employees as to the reasons for real and/or perceived roadblocks.'[37] Again though, the innovation team was a 'one-off' exercise. Once it had collected responses from the workforce on innovation at the workplace and made its recommendations, it was wound down by mine management, who hoped, somewhat unrealistically, that this would finally close the book on past attitudes.[38] Some of the recommendations of the team were, in their own way, quite insightful. Perhaps the most telling was a stated need for management to draw out the full connections between innovation practices and the performance of the profit-sharing plan for workers.[39] A more deliberate and planned move into the production politics of post-Fordism, as the next case illustrates, would have found this point to have been obvious.

The other concern that is often taken to be a crucial indicator in the transition out of Fordism is the so-called flexibility issue. An adaptable, multiskilled workforce is taken to be the sine qua non of economic survival in a more competitive, global economy. A regime of job control unionism, on the other hand, is viewed by many as the chief obstacle to attaining greater flexibility and competitiveness.[40] The search for flexibility at Central Canada has involved two components: adjustments to CWS job classes; and more recently in 1994, efforts to introduce a needs assessment program.

The existence of CWS at the mine would seem to militate against the type of workforce flexibility that is currently in vogue. Indeed some, including myself, would consider it to be the epitome of job control unionism. CWS was adopted at Central Canada in 1975 as part of that year's round of collective bargaining. This makes CCP the first site in our study to cede this form of job classification system to the local union.

It appears highly unlikely that the union would readily relinquish CWS in future rounds of bargaining. It is simply too important a lever to forgo. Nevertheless, it is management that ultimately controls employment levels at the mine. In the context of another managerial right, the ability to introduce technological change and the downsizing effects that have ensued from it, there has been a certain amount of de facto cross training, or enhanced flexibility. This has been reflected in the revision of certain CWS rates for specific jobs. That is, as work areas and responsibilities have been broadened in the wake of technological change and downsizing, specific functions have undergone CWS reevaluations and upgrading.

This has occurred in the hoist and shaft areas of the mine, where workers now do double duty in the guardhouse.[41] It has taken place in the mill where previously separate areas such as crushing, scrubbing, flotation, and drying have been combined into one job, compaction and screening have been merged into another, and crystallization and reagents have been united into a third.[42] Underground, since 1988, mine borer operators have been given new duties as 'worker coordinators' with organizational responsibilities for other crew members. Unlike the other sites in this study, where skilled trades people occupy the highest job classes, miner operators at CCP now hold this distinction, on account of their enhanced accountability.[43] These changes have coincided with a general decline in the ranks of the front-line supervisory staff, the foremen. Underground foremen have been thinned out by up to two-thirds since 1985.[44] Meanwhile, the traditional position of foremen has also been altered. At CCP, they are now called shift coordinators, and are seen by some to carry less authority than at one time.[45] In some areas, workers have been allowed to select the relief coordinators (or temporary foremen) on their own. Not surprisingly these alterations have received less-than-full endorsement from front-line staff, who have argued that the borer operators are not assuming the responsibilities for their crews and work areas as was intended by the changes. On account of this, underground foremen have come out against the job class and the associated rate that has been assigned to the coordinators.[46]

Thus, although saddled with CWS, management has attempted to increase its room for manoeuvrability by letting the size of the workforce slide (that is, downsizing), by combining the remaining jobs, and then by re-evaluating them, as prescribed by CWS. Amongst many, two of the side effects of this dynamic were drawn to our attention. The pursuit of leanness has had mitigating repercussions on the parallel apprenticeship program. With fewer workers, management has had less flexibility to follow through on the protocols of this agreement to the point where the union has come to consider the program to be inoperative.[47] And second, although individual jobs have been upgraded by means of CWS reappraisals, when they have been united with other functions, it *may* be the case that the new rates are still lower than the average rates that prevailed in an area prior to job merger.[48]

Another strategy for attaining flexibility is covered by an undertaking of more recent vintage, the Needs Assessment program. This has been introduced in the guise of a safety/training endeavor, but its implications have definitely spilled over into the realm of workforce flexibility. The program commences with a safety and training audit. Knowledge that workers ought to have in order to do their jobs safely is measured against existing training levels and any shortfalls are compensated for through additional training. In practice, however, this has entailed training on pieces of equipment and procedures that are not normally part of existing job descriptions. For example, members of the shaft crew and mechanics have received instruction on the operation of specific pieces of mobile equipment, such as scoop trams.[49] From the union's perspective, this is 'multiskilling' in its most pejorative sense. Cross training of this nature is viewed as contributing to leanness over the worksite as a whole, as well as to real skill dilution, or the 'jack of all trades, master of none' phenomenon.

If each operation in this study is readily definable by a certain outstanding trait, it is ambiguity that marks management strategies at CCP. At one and the same time, this is both a particularly clear and a fuzzy instance of social change in employment relations. Clear, in the sense that dissatisfaction on the part of management with the status quo has led to an incorporation of many of the elements that are constitutive of post-Fordist work relations. Performance sharing, employee stake/stock holder status, the generation of values and vision statements, climate surveys, recognition programs, and work restructuring around innovation teams are all indicative of post-Fordist nostrums. At the same time, the pursuit of these alternatives has been chaotic. Initiatives have been undertaken in the absence of clearly defined expectations. The existence of innovation and special project teams, for example, has been episodic. But perhaps the most striking instance of this is the belated identification of enterprise goals. Only after many of the above programs were put in place, were specific production objectives set. In this case, a 10 per cent reduction in the costs of production was identified, along with better utilization and productivity results from the new capital equipment that the company had installed.[50] Still, there does not seem to have been a clear effort to think through the relationships between the reforms that had been introduced over the preceding few years and specific production-related goals.

These observations are not intended to constitute a critique of managerial practices at this particular company. Rather, their purpose is to demonstrate that although the practices embody key elements of the new work relations, they have not been pursued to their full logical consequences. CCP represents more of a 'halfway house' in the transition out of classical Fordism. This will become clearer when it is compared with the last site in the study, Agrium Ltd.

Agrium and the Ethic of Continuous Improvement

The final case, Agrium Ltd. (previously Cominco Fertilizers), exhibits certain similarities to Central Canada Potash. Until 1993, the potash operation was part of a much larger corporate organization that specialized in the production of base metals and concentrates (lead, zinc, copper, molybdenum, and germanium), precious metals (gold and silver), as well as fertilizer products (potash, nitrate, ammonia, and urea). Up until 1993, when the fertilizer component of the company was spun off into a stand-alone unit, Cominco Fertilizers, metals production accounted for 75 per cent of corporate revenues and 83 per cent of total employment, while fertilizer production made up the remaining 25 per cent of company revenues and 14 per cent of gross employment.[51] As in the case of Noranda, then, the fertilizer and potash component of Cominco's business was distinctly secondary to other concerns. Unlike Noranda, on the other hand, Cominco demonstrated less overall diversification. Strictly speaking, Cominco is not a natural resource company, as is Noranda, but a mining/smelting conglomerate. In addition to the huge lead-zinc smelter that the company owns and operates in Trail, British Columbia, it is also the sole or majority owner of another seven mines, while enjoying important minority shareholder status in an additional three mining operations in Mexico, Australia, and Spain. The bulk of these properties are lead and zinc operations.

In contrast, the fertilizer portion of Cominco's business was far more comprehensive than Noranda's. Cominco, prior to the spin-off of its agro-industrial assets into an independent entity, held six properties that produced nitrogen, phosphate, and potash. It was this level of integration that made the establishment of a separate Cominco Fertilizers a real possibility. This took place in 1993 and included the one potash mine that Cominco possessed. In 1995, to delineate its new independence, Cominco Fertilizers changed its name to Agrium Ltd.[52] At the time of the study, this new entity included four nitrogen plants, two in Canada and two in the US, which produced anhydrous ammonia and ammonium nitrate; the one Saskatchewan potash mine; and the agro-market retailing outfit, Crop Production Services Inc. Only two of these operations were unionized, while the marketing wing of the company was set up as local franchised businesses. Overall the bulk of Agrium's markets have been North American. This has included its potash business, where 78 per cent of its product has been sold for North American crop production.[53]

Keeping these similarities and differences with Central Canada Potash in mind, social change in work relations at the potash division of Agrium was born on the wings of a perceived crisis, just as it was at CCP. At different times this was described either as a crisis of declining market share, especially in the North American theatre, or alternatively, as a profitability crisis occasioned by rising production costs.[54] These issues were compounded by the company's initial reaction to them, which was to move into a regime of strict scientific management in the latter 1980s.[55] This was brought to Agrium through the auspices of a consulting firm, Proudfoot Consultants, which specialized in these techniques. Time and motion studies were conducted throughout the operation, leading to a significant downsizing exercise and increased monitoring of those workers who remained. In total, sixty-two non-managerial jobs were targeted for redundancy, while monitoring equipment was installed on the continuous mining macnines to keep stricter track of work effort.[56] Tellingly, there was little organized resistance to this assault. In spite of unofficial work slowdowns, the local union remained more preoccupied with the question of the role of seniority and its processing in the pending lay-offs than with challenging the consultant's report.[57] Significantly, this action came *after* the worst years in the industry had passed, which made it all the more difficult for the workforce to excuse, or to forget. Thus, just as business was beginning to rebound, workers at Agrium were hit with large-scale job losses and a 'get tough' approach to running the business. The implications of this legacy are still very much a part of the collective memory at the mine.[58]

Ironically, most of the job losses had to be rescinded about a year and a half after they had been issued. The scientific managers had gotten it wrong and the cutting had been carried out too deeply for the plant's own viability. In the interim, operations became badly rundown, overtime payouts expanded exponentially, and morale plummeted.[59] It was in this context that the company made a dramatic volte-face, towards the adoption of new employment relations. These have gone by different names and have had a variety of program components, but overall they have been characterized by two main traits. First, they have emanated out of head office as top-down initiatives. Second, they have inevitably been grounded in an ethic of continuous improvement that has provided the guiding theme at Agrium since 1990. As such, the restructuring of work relationships at Agrium has had a more consistent, strategic application than has been the case at CCP. Consequently, Agrium has also traveled further down the road of post-Fordism than is obvious at the other sites included in the study.

In the wake of the scientific management debacle, Agrium officials realized that they had a significant morale problem on their hands, on top of the market share and profitability issues that were previously alluded to. These issues subsequently came to occupy human resource management at the company in some rather unique ways. The first response, which like all the others must be viewed as *both* an effort to undo the damaged visited upon the operation by the experiment in Taylorization and an attempt to address the issues of production costs and profitability, was the release of a new program, entitled 'The Road to 4X,' or more simply '4X.' The title of this project hails from the goals that it embodied, namely, 'to achieve dramatic improvement in our capacity to create increasing value for customers, shareholders and employees.'[60] This was to entail increasing company earnings and shareholder profits by a factor of four (hence 4X) in the seven-year period running from 1990 to 1997. Four new initiatives were taken to accomplish this including induction of a cultural shift at the company, the adoption of an integrated profit focus across operations, the formation of strategic alliances with customers and other significant actors, and the creation of a new emphasis on research and development. Each of these new strategic undertakings was intended to lead to the attainment of a global presence for the company. For our purposes, only the first is consequential. The cultural shift referred to entailed a transition to employee involvement in the labour process, as well as the adoption of customer, growth, and profit orientations on the part of the workforce. For company planners, this could only be accomplished by achieving 'a work environment in which employees are challenged, have opportunities for personal growth, are living our core values, and are focused on winning (achieving) these goals.'[61]

In the minds of head office planners, this change in *weltanschauung* could only come about through intentional workplace reform. The vision would embody a movement away from detailed job descriptions and guidelines, to formats where workers would have the latitude to decide how to tackle a broad array of jobs, and supervisors would be left free to plan and facilitate.[62] In practice, this would entail de-emphasizing seniority, upgrading levels of workforce training, and creating, in the words of one document, a company 'learning experience.' Above all, work should be redirected towards the continuous improvement of all operations, through the utilization of semi-autonomous work teams. As enshrined in the company's new 1990 mission statement, 'We believe people working together, combining their strengths, create a more powerful force than individuals working alone.' Employee recognition programs and the establishment of mechanisms to lend a 'pride of ownership' in innovative ideas were intended to add substance to such declarations.[63] Thus, as prescribed in one corporate document, 'Managers make every effort to encourage workers to examine their jobs and work areas and think of ways to improve them, no matter how primitive they may be.'[64]

In its pursuit of a less bureaucratic, more participatory work culture, Agrium possessed one major advantage, as well as one significant liability. The latter, as already mentioned, was the legacy of sour relations and mistrust that resulted from the company's initial response to workplace reordering, the turn towards Taylorism. This bequeathal would continue to haunt employment relations at Agrium long after the original program had been abandoned.

Over and against this history, however, was an important structural aspect of work relations at the company. Alone amongst all of the sites included in the study, Agrium has not been saddled with a CWS job classification system, or a managerial proxy for such, as was found at Key Lake. For reasons of their own, when other mines were adopting the job control measures of CWS, Agrium

TABLE 4.4
Agrium occupational structure and wage structure

Mine/mill operators	Maintenance trades	Rates (1994)
	Journey tradesperson	$22.21
Journey operator	Tradesperson (non-ticketed)	$21.29
Operator A	Tradesperson A	$19.44
Operator B	Tradesperson B	$17.64
Operator C	Tradesperson C	$15.83
Operator D	Tradesperson D	$14.10

Source: Memorandum of Agreement, Agrium Fertilizers and United Steelworkers of American, Local 7552.

workers chose not to follow suit.[65] As a result, Agrium has had a completely novel occupational structure. As opposed to the multiple lines of departmental progression and numerous job classification levels that one finds at the other sites, at Agrium there have been only two tracks that culminate in two final job rates. For the non-apprenticed occupations, workers enter into training operator positions. After five years, the employee has been trained on every piece of equipment and work area in the mine, or the mill, and is given 'journey operator' status, along with the top wage rate for operators, regardless of the particular job that is being performed. Analogously, trades workers pass through a four-year apprenticeship program, attaining journey trades status in their fifth year. With the exception of recent hires (that is, those with less than five years' seniority or Operator and Trades positions A through D in Table 4.4), there are in effect, only two job classifications and two corresponding wage rates at Agrium – journey operator and journey trade positions. Between the two, as indicated in Table 4.4, there is less than a dollar differential ($.92 per hour), in favour of the ticketed trades.

This exceedingly flat occupational/pay grid, which has entailed comparatively higher labour costs for the company, was *originally* promoted as a means of reducing labour turnover and maintaining a skilled workforce.[66] For the same reason (that is, the favourable wage structure), workers opted to retain it when given the choice between the status quo and a CWS system, in the early 1980s. The significance, however, is that such an occupational grid *invites* greater flexibility in the allocation and use of labour in the labour process. From management's perspective, the company was already paying the opportunity costs for such potential flexibility.[67] Following the demise of scientific management at Agrium, management began to become aware of just what an opportunity it had inherited. From this point on, a dedicated effort was expended to take advantage of it.

The 4X initiative at Agrium had many of the hallmarks of Total Quality Management (TQM) built into it. Management was instructed to treat each division of the new company as a stand-alone profit centre.[68] Optimal profit centres, (customers, regions, et cetera) were targeted for special attention under the credo that 'Last year's performance is always unacceptable.'[69] In no uncertain terms, the company was reinventing its image, as a 'service company focused on the changing needs of our customers and their customers – not only today, but also tomorrow,' through dedication to value-added processes.[70]

Launching TQM at Agrium commenced with an employee climate survey and the establishment of a requisite committee structure to oversee the operation of the new paradigm. Periodic workforce surveys were viewed as an important part of the project. They would provide a means of measuring the cultural shift that was supposed to be taking place through the development of a team-oriented, problem-solving environment.

Under the 4X program, a joint steering committee composed of management and union representatives was established along with four subcommittees, consisting of employee involvement, job security, health and safety, and employee recognition.[71] The first of these, the involvement committee, was given the mission of promoting an ethic of continuous improvement in the workplace, while the latter body was charged with drawing up a gainsharing plan in support of the continuous improvement effort. Administration of both initiatives was the function of the joint steering committee. This group also dispensed team charters and vetted work team proposals as in Figure 4.1, which illustrates the model that the steering committee was supposed to employ. Those proposals bearing upon traditional managerial rights would be referred to company management for final disposal, while suggestions that might have an impact on employee rights would be routed to the union for ultimate disposition. As can be seen in the company's own roles and responsibilities chart (Table 4.5), however, the joint steering committee was to play a largely facilitative role to management's business plans, lining up support for continuous improvement and other managerial decisions over the allocation of corporate resources amongst the other actors. As indicated in Table 4.5, management continued to hold principal, or sole, responsibility in key areas of decision making, such as setting a corporate direction, establishing business priorities, and allocating resources. The joint steering committee, specific project teams, and the union, on the other hand, were to assume a distinctly junior role, being informed, or at most, consulted about such plans. Union involvement was to be mainly limited to a support capacity, of getting members involved in the continuous improvement initiative.

FIGURE 4.1
Model for the functioning of continuous improvement program at Agrium

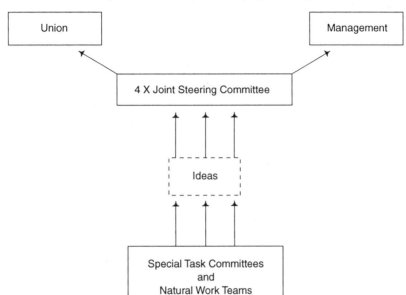

Source: Agrium Joint Steering Committee Minutes, 14 February 1992.

The chartering of natural work and special task-force teams was a major responsibility for the new joint steering committee. In the company's view, a work team was to be 'a chartered group of employees who are responsible for a "whole" work process or segment that delivers a product or service to an internal or external group.'[72] Each team charter would outline the resources available to the team for the realization of goals, including money, time, training, and allocated authority. According to corporate policy, 'Team members may be expected, at various times, to perform all the jobs of the team within the boundaries of their particular expertise.'[73] Basically, the theoretical expectation was that team members would manage the group effort on their own, while management would set a strategic direction for continuous improvement efforts around the site. Within this framework, it was hoped that work teams and their members would develop a sense of 'ownership' in their work and in ideas that would improve it. To give this effect, 'value idea forms' were freely available to members of the workforce, and managers were encouraged to keep account of spe-

TABLE 4.5
Designated roles and responsibilities in Agrium continuous improvement project

Roles and responsibilites	Stakeholder				
	Senior management	Joint steering committee	Foremen	Project teams	Union
Vision/direction					
business	PR	IN	IN	–	IN
workplace	PR	IN	IN	–	IN
Establish priorities					
business	PR	IN	IN	–	IN
workplace	PR	CO	JR	–	JR
Communicate					
business & workplace	PR	JR	JR	–	JR
Enrol					
get people involved	PR	JR	JR	–	PR
Allocate resources					
time	PR	CO	JR	JR	IN
money	PR	IN	JR	CO	IN
training	PR	CO	CO	CO	CO
support	PR	PR	JR	JR	PR
information/knowledge	PR	CO	JR	IN	CO
people	PR	CO	JR	JR	IN
Remove barriers	PR/JR	JR	JR	–	PR/JR
Manage/create policies & systems					
manage	IN/CO	IN/CO	PR	IN/CO	IN/CO
create	JR	JR	CO	CO	JR
Training & coaching					
manage	–	–	IN	–	–
improve	CO/IN	IN	CO/IN	PR	CO/IN
Monitor progress					
con't improvement	JR	PR	JR	PR	JR
Assess & communicate key results					
business	PR	JR	JR	JR	JR
workplace	PR	JR	JR	JR	JR

Key: PR Prime responsibility; JR Joint responsibility; CO Consulted; IN Informed;
AC Accountable; – No involvement.
Source: Agrium, Roles and Responsibilities Chart, 14 April 1993, mimeograph.

cific value-added suggestions, along with the numbers of proposals received per month.[74]

Another dimension to the adoption of a team-oriented approach is to be found in the advocacy of a new relationship with the union at Agrium. Union participation was built into the designated roles and responsibilities of the continuous improvement project. To compliment this, the company advocated supplanting traditional, adversarial relations with a forum for joint problem-solving. At this level, management advocated use of the side by side problem-solving approach, as developed by the Harvard Negotiation Project, and its strategies of 'getting to yes.'[75] In this paradigm, both management and union are confronted with the common task of finding an agreement that does not leave joint gains 'on the table.' In other words, the object is to conclude an agreement with a situation that expresses Pareto optimality. Accordingly, the art of negotiation is to resolve *interests*, through the exploration of a range of options, rather than to defend pre-set *positions*. Together, management and union are charged with brainstorming for options that resolve the interests of both parties.

It is always difficult to discern how widely practiced, or how influential a new managerial paradigm is in reality. Indeed, this has been one of the central issues in the labour process debate, as for example, in arguments over the efficacy of scientific management in history as opposed to in text.[76] We are faced with a similar problem here. To what extent was the new vision at Agrium, as outlined above, a significant influence on the organization of work? To what extent was it a 'paper tiger'?

As a first approximation it can be reported that a joint steering committee and the four supporting subcommittees were established and met on a weekly basis. Joint lists of expectations to come out of 4X were established by these committees.[77] In turn, the subcommittees developed approaches to employment security, as well as plans for a gainsharing program.[78] Additionally, a number of special task-force teams were founded at this time and chartered by the joint steering committee. These included a dust and fines reduction team in the underground mining department, special project teams in the mill, and a product-handling team in loadout.[79] According to the model of team charters drawn up by the joint steering committee, such teams were to scope their own work, schedule team meetings and work allocation amongst group members at the start of shifts, document their work, and have input into the allocation of moneys affecting their work.[80] Of course, one should not jump to conclusions here, especially when it is realized that throughout the early history of 4X, ongoing debate was recorded over the role that supervisors would play in the team approach. In particular, worker members of the joint steering committee expressed concern that with supervisors as automatic members of the work teams, the real seat of control would not change.[81] In the end, supervisors were deemed to be members of the natural work teams, but not necessarily team facilitators, while supervisory membership on the special task/process improvement teams remained optional.[82]

To return to the issue of significance, three other indicators have a bearing on just how much importance should be attached to these developments. First, management invested considerable sums in the cultural shift that they were hoping to invoke. By company estimates, $300,000 were expended in lost wages for workers attending steering committee, subcommittee and work-team meetings. Task team members were paid overtime rates for any participation in team meetings that occurred outside of official work time.[83] These sums do not include the hefty expenditures that were also made for management and team training seminars on the ways and means of continuous improvement, or for gainsharing program consultants. Obviously, then, management was taking the movement into continuous improvement very seriously indeed. A partial by-product of these pursuits was the minimization of lay-offs at the mine. That is, management made the decision to run at lower throughputs than was possible, so that workers could participate in the 4X initiative. In so doing, management was consciously attempting to avoid any lay-offs over the course of the production cycle.[84]

Second, some evidence is available from interviews and information gathering conducted by the joint steering committee itself, pertaining to the effectiveness of its initiative. As part of its mandate, this overseeing body was to collect information on the diffusion of continuous improvement across the site. The steering committee was especially interested in the spread of the team concept and the need for additional training in this approach, as measured against the current adequacies of team functioning. In recorded interviews, conducted by the joint steering committee of subcommittee and special task-force team members, participants voiced support for the team concept and the manner in which the existing teams were operating.[85] In other words, there was a basic familiarity on the part of Agrium's workforce with such concepts as teamwork and continuous improvement. These impressionistic findings are also supported by our own data collection. When asked if they were familiar with a number of new organizational paradigms, higher proportions of Agrium workers indicated having knowledge about quality circles, TQM, and Quality of Working Life (QWL) programs respectively, than was the case for the workforce samples at the other mines.

Third, after some initial reluctance, the local union did get on board the initiative. And although union support continued to wax and wane, depending

TABLE 4.6
Agrium portrayal of old and new business pardigms

Old (Fordism)	New (Continuous improvement)
Quality drags productivity	Quality drives productivity
Doers separated from thinkers	Doers must be thinkers
Assets = thinkers	Assets = people
Profit is number one	Customers are number one
Hierarchical focus	Network/web focus
Measurement to judge results	Measurement to aid improvement
Industrial society focus	Information society focus
National economy focus	Global focus

Source: Training seminar given to Agrium employees by Tennessee Associates
International, 1992.

upon local politics and the current overall state of labour/management relations at the mine, a letter of understanding was entered into with the company, in support of the 4X undertaking.[86] As part of this initiative, joint problem-solving techniques, embodying the non-adversarial relations of 'getting to yes,' were utilized in two rounds of contract renewal. In the first of these, the meetings were open to outside observers, and according to the company director of human resource management, it would have been difficult for a stranger to discern who was representing management and who was there on behalf of the union.[87]

If anything, these forays into the new industrial relations would be extended in the succeeding years. Midway through 1992, members of the joint steering committee were enrolled in a training seminar on the methods of continuous improvement. This signified a further clarification in emphasis, as was denoted in the change of name that the 4X steering committee underwent at about the same time, becoming the continuous improvement joint steering committee. Advocates of continuous improvement recommend that employees be 'empowered to make changes to improve processes and receive rewards and recognition consistent with continuous improvement.'[88] This is to be achieved through the chartering of both natural work and process improvement teams by a steering committee, which 'own(s) the implementation process.' As implied in the team concept of work organization, employees who are cross trained in different facets of the operation are better suited to identify and remedy process problems.

For subscribers to the new regime, continuous improvement represents nothing less than a paradigm shift in the way that production is carried out. Table 4.6, taken from the training program that officials at Agrium used, highlights what are put forward as the main differences between traditional (that is, Fordist) production models and continuous improvement ethics. According to the guidelines that the company followed, 'The key to managing for continuous improvement is to manage process,' with the object being the elimination of non-random sources of variation in both the labour process and its outcomes. Within the labour process, unacceptable variation is constituted by anything that interrupts the smooth flow of work processes. Physical bottlenecks, high ratios of work in process, or inventory, and idle human resources are all accepted indicators of suboptimal performance. Thus, continuous improvement recommends the study of labour processes from 'a time perspective.'[89] Here, the object is the quite straightforward reduction of cycle times. The training program that Agrium utilized, stated: 'One of the biggest sources of waste in any organization is inefficient use of time. A basic premise in process management is that 'time is money' and cycle time is a key measure of process efficiency.'[90]

Statistical process analysis is the preferred method to be used in distinguishing normal process variation from excessive variation caused by specific problems. Such problems are identified as the special focus of the work teams that are constituted under continuous improvement. In other words, teams are responsible for the identification and description of process problems, the verification of the cause(s) of unsatisfactory variation, the implementation and evaluation of solutions, and the standardization of the superior process. As part of this mandate, teams are given special responsibility for conducting 'value added-assessments,' the object being the minimization of essential non-value-added work and the elimination of non-essential, non-value-added activity. This entails the standardizing of work procedures, so that best practices can be benchmarked and used as the standard by which all workers are judged, while at the same time 'error proofing' the labour process against possible defects. As the program that Agrium subscribed to advised, continuous improvement involves 'An ongoing investigation and learning experience that ensures that the best processes are uncovered, analyzed, adopted and implemented. ... Find the best practices and steal shamelessly.'[91]

The continuous improvement ethic extends beyond the realm of production, but in such a manner as to feed back into it. Markets are brought back into the centre of the labour process as it were. Thus, customers are not only the final purchasers of output, but also fellow employees, who happen to be working downstream within the same labour process. At the same time, a renewed focus is placed upon customer satisfaction and tying it together with process improvement in production. Thus, 'Customer's satisfaction is directly correlated with an organization's ability to focus, define, analyze and improve processes.'[92]

Research into customer needs ought to become a basic part of the business, and such knowledge used in product design requirements, product features, and process steps. For these purposes, customer input through periodic surveys, or what is termed 'voice of customer analysis,' is essential. This is to be used to pass beyond basic quality and performance quality levels to the third tier of what pundits refer to as 'excitement quality,' that is, product standards that exceed the client's expectations. A moment's reflection on these points begins to explain why Agrium was redefining itself as a 'service company,' dedicated to research and markets. And indeed, in this very primary industry, more individuals are still employed in the company's marketing and service subsidiary than in any other activity.

Despite the role ascribed to teams in norming work processes, the continuous improvement approach makes it clear that overall strategy is management led. Corporate planners have ultimate responsibility for approving projects, assigning them to the most appropriate team, and providing the team(s) with the necessary resources, including the legitimating team charters to complete the projects that have been approved.[93] As improvements in process are registered, they come to represent the new benchmarks for further improvement in a world where 'last year's performance is always unacceptable.' At the same time, an overall sense of direction is provided by senior management, focusing on strategic initiatives and long-term goals. Certainly, the original 4X program would fall within this category. With its specified production and profit targets, a functioning joint steering committee, and the process task teams that were struck, the strategy of continuous improvement was already well underway at Agrium.

To underscore the point, the formal letter of understanding that was signed with the union stated that both parties undertook 'To create an environment where all employees feel secure in participating in processes that *continuously improve* the Cominco Fertilizers' Potash Organization and Operation ... To enhance our customer relationships and our competitive position in the market place [my emphasis].'[94] These desideratum were to be realized through a commitment by both company and union 'to participate in consensus process (solutions for mutual gain).'[95] Additional substance to this letter appeared in the form of a draft agreement on Proposed Guidelines For Continuous Improvement that was made at the end of 1993. This statement of understanding between the local union and Agrium provided additional local site specifications to the original 4X project, which had now been solidified as a continuous improvement program. It commenced with a recognition that long-term viability was dependent upon 'more effective use of the collective knowledge and skills of all employees by constant attention to finding ways to improve the operation.'[96]

Under the guidelines, continuous improvement would be directed towards the realization of efficiency, quality, safety, employment security, recognition, and commitment goals. Certain of these objectives were spelt out in considerably greater deal than others. For example, in operationalizing efficiency norms, the statement concurs in the belief that 'the most appropriate means of creating optimum effectiveness and efficiency is to create a structure in which ALL employees can contribute to their maximum potential, willingly, to achieve jointly developed common goals and objectives. The best structure for this is a TEAM environment. Natural work teams [normal work crews] and special task teams, [short term, specific project, cross functional groups] will operate under the following recommended guidelines.'[97] To give this effect, the guidelines on continuous improvement specified that work teams ought to be limited to no more than eight workers per group. Membership is voluntary, although supervisors are automatically part of the natural work teams. Team leaders are selected by the group as a whole, and do not necessarily have to come from the ranks of front-line management. Meetings are to be work related and decisions are to be value added in nature.

With respect to other issues, such as those pertaining to job security and work safety, the guidelines were considerably less detailed. Nevertheless the two items on which the future of continuous improvement at Agrium would depend were those relating to job security and employee recognition. With respect to job security, there is an obvious tension between continuous improvement and employment continuance. Indeed, continuous improvement could be taken as encapsulating the very notion of our theme, 'more with less.' To get around this issue, the proposed protocol on continuous improvement stated: 'Job security is understood to be the singular most important concern of employees. Although there can be no absolute guarantee of jobs for anyone in the competitive world of today, it is recognized that IMPROVEMENTS justify, and provide the means of sheltering employees much more effectively from involuntary loss of jobs. Should IMPROVEMENTS provide the opportunity to do all required work with fewer people, the following process would be implemented IN THE SEQUENCE LISTED to deal with people impacted by the elimination of a job.'[98] The document then goes on to enumerate the options that would be followed, including, in the order in which they would be invoked, job transfers, retraining, attrition, early retirement, job sharing, and finally enhanced severance.

If competitive position and the implications of this for individual job security is the stick that lies behind continuous improvement, then employee recognition represents the other side of the coin, the sweetener, or inducement for approval of the continuous improvement ethic. In particular, monetary recognition ought to be built into any initiative 'to return to the employees, a fair portion of saving

resulting from improvements generated on site.'[99] Such worker recognition programs are to 'Provide the means by which employee created ideas, which
improve potash production and/or reduce operating costs, can be measured and
rewarded.'[100] As much as anything else, this is what is taken to give continuous
improvement its win/win quality, or as baldly concluded in the document:
'Let's face it; there is no need to hide ideas – everyone will gain from them and
no one gains if they aren't brought forward!'[101]

The worker recognition scheme contained in the statement of understanding
between Agrium and the union was intimately tied to the continuous improvement project. For this reason, the proposal *was not* a profit-sharing plan. At
Agrium, management viewed profit sharing as being too broad, in the sense that
final payouts were contingent upon too many factors that lay beyond the realm
of the work-effort bargain, including variables such as the state of product markets.[102] Instead, management wanted recognition to be tied strictly to fixed production costs and increased efficiencies, as measured in declining costs, or what
was referred to as 'legitimate gains.' The use of fixed production costs as a
measure of process improvement offered other advantages as well. Information
on this topic could easily be propagated throughout work areas and around the
operation, thereby 'educating' workers to the 'realities of the business.'[103]
Production costs also presented a readily accessible moving target, one on
which a definition of continuous improvement could easily be hung. With these
rationales foremost, the employee recognition program at Agrium assumed the
form of a gainsharing plan. The original proposal allowed for a joint recognition committee composed of equal numbers of union and salaried staff representatives. Savings accruing through continuous improvement were to be split
on a sixty:forty ratio, skewed in favour of the company in recognition of the
'need for payback on investments of equipment and materials.'[104]

A final piece in the continuous improvement project at Agrium was to be
found in the proposal for a specific training program that would complement
the overall effort. The issue of training has already been alluded to in several of
the contexts associated with continuous improvement, most notably in references to the centrality of cross-functional work teams. In order to bolster this
objective, the company proposed utilizing a competency development program
as a critical instrument in the realization of continuous improvement. With the
precepts of this undertaking, local design teams of front-line workers would
author specific job profiles, complete with levels of required competency for
each function on site.

Insight into the nature of training in a continuous improvement regime is
afforded by the very notion of competency that is addressed in the program.
Thus, competencies are defined as 'Knowledge, skills, abilities, talents and

behaviours that if effectively deployed, result in increased competitive advantage and improved ability to create economic value.'[105] Nine areas of training were identified by the company. The list is indicative of how employers frame the issues of training and skill in a post-Fordist environment. It includes the following items, which were to constitute the centre of the competency development program:

1　Knowledge of the business
2　Knowledge of continuous improvement
3　Planning and organizational skills
4　Leadership skills
5　Customer knowledge
6　Ability to develop effective working relationships
7　Analysis and decision-making abilities
8　Communications skills
9　Ability to encourage people development in self and others

In each instance, one out of four competency ranges, (learning, doing, sharing, or shaping) is deemed applicable, depending upon the job profile that has been created for each position. Thus, for any given job and with respect to a given focus, a learning knowledge may be defined as appropriate, while on another dimension of the range of competencies, doing or sharing abilities may be required.

Operationalizing the competency development program provided the other notable feature for this aspect of continuous improvement, and indeed, should be viewed as an offshoot of the latter. The whole initiative assumed some form of 'skills inventory' to be conducted through individual worker assessments. In this manner, existing levels of competency could be measured against the desired job profiles and any shortcomings could be remedied with the appropriate level of training, in customized 'training action plans.' To deal with the issue of assessment, competency development recommended a form of peer review and complementary forms of reward. Envisaged by company officials were evaluation teams composed of the employee's supervisor and equal numbers of managerial and employee selected assessors.[106] Also alluded to, but not substantially developed, were plans to further link competency development with compensation. This could include an element of merit pay for workers who demonstrated the right competencies in an appropriate fashion.

As emphasized by the training coordinator for competency development, who just happened to be the former local union president, the initiative was principally concerned with the development of so-called soft skills; that is, basic 'people skills' that would allow individuals to manage their own work more satisfactorily, while acclimatizing to the new culture of continuous improvement.[107] Indeed, if one examines the range of training foci that are addressed in the project, tangible or 'hard' skills can possibly be associated with only four of the nine competency areas listed above. *Some* knowledge of statistical process control may be required for proficiency in continuous improvement, while varying degrees of analytical or symbolic acumen are most likely required for a demonstration of communications skills, for conducting information analysis, and for exhibiting a basic knowledge of the business. But even here, and more so in the other areas of defined competence, what is being sought is more an intimate knowledge *and acceptance* of the company's culture than anything else. This was noticed and remarked upon at the mandatory information seminars that were held for all employees on site, at the time the program was rolled out. Here, the project was branded by participants as, among other things, 'a waste of time,' 'Japanese-style managerial regimentation,' and 'cult-like.'[108] In both meetings that I attended, the training facilitator soon had to depart from the prepared presentation of overhead visuals and text. One meeting used less than an hour of the allotted three-hour period, breaking up in an atmosphere of sullen hostility, with employees preferring to return to work rather than sit through the complete seminar. Uncomplimentary cartoons, comparing competency development and the training coordinator with the consulting firm that had formerly brought scientific management to the site also appeared on bulletin boards during the seminars.[109]

These latter observations should serve once again, as a warning against reifying managerial intentions into labour history. At Agrium, a holistic design for continuous improvement was put into motion. But this was not the end of the matter. Union and worker responses were also critical in determining what would happen to the continuous improvement initiative. The union position with respect to the new apparatuses of production was uncertain, at best. Elements of mistrust, 'strategic opportunism,' caution, local union politics, and the larger relations between the local union and the national office operated to varying degrees, at different moments.

Mistrust was occasioned by previous misadventures with the company, and in particular with the downsizing fiasco of the scientific management endeavour. The legacy of the Proudfoot program into mainstream Taylorism left the union particularly sensitive to any moves that might further reduce employment levels at the mine. Owing to the failed experiment in Taylorization, the union was now considerably more vigilant than it had previously been. Hence, throughout the continuous improvement initiative, job security remained a sticking point with the union and its level of involvement.

Simultaneously, however, the local union was concerned that it might be marginalized in ongoing company affairs should it simply refuse to participate in the new corporate initiatives. Simply saying 'no, not interested' was perceived at both local and national levels as being inadequate.[110] Echoing national policy, local union officials felt that they should serve on continuous improvement committees, attend management workshops on the subject, and take part in worker training sessions, largely as part of an ongoing monitoring activity. This is what can be referred to as 'strategic opportunism.' Participation by the local union allowed executive members both to keep tabs on what the company was up to and to gauge how it was going over with the membership, thereby providing the union with more input with which to draft responses. This, for example, was precisely why the local president attended every crew meeting that was associated with the introduction of the company's competency development program.[111]

The notion of 'strategic opportunism' is also used in a second, broader sense. Here, I wish to reference an attitude that can be encapsulated in the notion that 'if the company wants to pay us for what we are already doing and call it something else (continuous improvement, gainsharing, etc.), then so be it.'[112] At the same time, this opportunism was always qualified in the following fashion. Any accretion in earnings through continuous improvement could not be 'blood money,' that is gainshares could not be produced or accepted at the expense of other members' jobs.[113] This qualification was what made the local's opportunism strategic.

These final two factors, which were considerations in the union's responses to continuous improvement, bear on both the politics of the local and its relationship with the national union and its representatives. By some measures at least, the local union at Agrium was a moderate one. This is best signified in its initial cooperation with the Proudfoot study, an event that would have been far less likely, indeed unthinkable, at some of the other sites included in the study. Following Proudfoot, the union was at first reluctant to get into a new venture with the company. Nevertheless, union representatives to the 4X steering committee were elected with instructions to oppose any undertakings that would be injurious to the membership. Interestingly, a sum of money was also set aside to pay for lost time on the part of the worker representatives to 4X, when they were caucusing separately as the union representatives on the committee.[114] Succeeding monthly meetings also featured regular reports on the 4X initiative, as well as critical discussion pertaining to it.[115] During this period, the local also authored its own strategic plan, putting forward a vision for its future and the evolving relationship with the company. This document appeared to buy into elements of the new managerial discourse, as when it stated: 'Our Local

Union's vision is that we will be recognized as an innovative, knowledgeable and respected Union team ... Collectively we will increase and share in the wealth of *our* company [my emphasis].'[116] In order to achieve this, the local set out as one of its principal objectives the need 'To build a productive relationship with the company which increases our ability to assist the company's long term wealth creation and thus our share of this wealth and job security.'[117]

Acceptance of continuous improvement along with the joint, side by side problem-solving approach were advanced as strategies through which to undertake these objectives.[118] These items were formalized, shortly thereafter, in a letter of understanding with the company on continuous improvement and in the acceptance of the non-adversarial problem-solving format, for the settlement of all non-monetary issues in dealing with the company – the 'getting to yes' technique already referred to.[119]

Correspondence with the national office indicates that there was less enthusiasm outside of the local for the new industrial relations climate at Agrium. Commenting upon the local's strategic plan, the National Director felt compelled to caution local leadership that 'it is always important for the local to constantly remind itself that working together with the company is a strategy; not a goal in itself.'[120] In this vein, the local executive was reminded that it was incumbent to clearly distinguish between the union's goals and the company's goals, as well as to educate the membership to this effect. There were also differences on more concrete issues. For instance, national union policy on gain-sharing urged that it be incorporated into the collective agreement and that workers take home 60 to 70 per cent of any registered cost savings.[121] This was just about the opposite of what the Agrium employee recognition subcommittee had proposed. Additionally, as in the national office's response to the Central Canada Potash performance-sharing plan, concern existed over the possibility of companies unilaterally canceling such plans after they had captured employee ideas and improvements.

These admonitions were to prove prescient, as the local first took a step back from continuous improvement and then, subsequently, withdrew from it altogether. Initially, the atmosphere was soured by spring contract negotiations in 1993. These were the first talks to use the joint problem-solving approach. Local management and the union did in fact reach a consensus on a new, two-year agreement, but much to everyone's surprise, this was vetoed by Agrium's newly created board of directors.[122] This setback was to create disenchantment within the local, which felt that cooperation had been reciprocated with manipulation, demonstrated by the re-emergence of unilateralist tendencies at corporate head office.[123] This event was to serve as a poignant reminder of where power truly lay.

Additional grievances were added to the following year's round of bargaining. Most important, Agrium announced a new round of redundancies, which, although limited in extent, placed the issue of job security and its relationship to continuous improvement, front and centre.[124] This seemed to catch the local off guard, while confirming the missives of the national organization. The local responded with a petition of its own, threatening to withdraw from continuous improvement should lay-offs proceed. In the end, the company scaled back the numbers to be let go, and sufficient numbers stepped forward to take early retirement or the voluntary severance package to temporarily defuse the situation.[125] Still, examining the different 'spins' that were placed upon this episode by management and the union is instructive.

For the company, the fact that there were no involuntary severances was proof that the new cooperative relationship embodied in continuous improvement was working.[126] For the union, on the other hand, even the threat of lay-offs was indicative of what the company was really after. The situation was made more complex for the union, by virtue of the fact that Agrium did not possess the standard CWS progression system, which is usually used to regulate bumping. In the event of a lay-off, it was unclear whether seniority would only be operative within separate departments, or whether it would take effect in a cross-departmental manner. From the union's perspective such ambiguities could not be allowed to persist.[127] The renewal of the collective agreement would provide an opportunity to address this and several other outstanding issues associated with the continuous improvement philosophy.

Most important for the union was an agreement on employment security. On this issue, and as a means of preserving the commitment to continuous improvement, the company offered a two-year moratorium on any job loss that might be connected with the program.[128] This was insufficient for the union and its members.[129] As a result the letter of understanding on continuous improvement, as well as the statement of understanding on guidelines for continuous improvement, were revoked by the union following a one day legal strike over the renewal of the collective agreement. As highlighted by a member of the union's bargaining committee during a presentation at the ratification meeting for the new contract, continuous improvement had lost credibility with the membership when the company unilaterally announced the lay-offs a few months earlier.[130] In spite of the fact that all job exits would eventually be voluntary and that the company took this as a sign of the success for its initiatives, Agrium workers had different perceptions. They had a four-year taste of continuous improvement and obviously retained serious afterthoughts.

Does this mean that continuous improvement is dead and that Agrium will revert to more traditional Fordist industrial relations? There are several reasons for doubting this. As already observed, the company has a non-Fordist occupational structure. It is unlikely that the current system of five-year journey operator and trades positions will be transformed into a more conventional job classification system, especially as none of the actors deem this to be in their interests. But, as long as the current broad-banded occupational structure is maintained, it will invite further initiatives into flexible production. As company officials readily admit, since they are already paying the price for a multi-trained workforce, they ought to be receiving some of the benefits associated with such. Clearly, they would like to see more flexibility in job assignments and movement around the site.[131]

Follow-up developments, after the short work stoppage, indicated that management was prepared to proceed on these premises. Thus, even after the formal withdrawal from continuous improvement, management approached the union for agreement on a gainsharing plan that would provide the focal point for improvement efforts on site.[132] Here, the intention was to 'Develop a simple, easy to use, systematic approach to regularly capture and implement employees' ideas for productivity improvements.' To bring the connection between effort and reward home, Agrium's consultants recommended that gainshare payouts be 'equitable, not equal in manner,' so that more efficient units within the workforce would realize proportionately greater rewards.[133] In this fashion, according to the consultants, 'The potential rewards [could] be of significant magnitude to change the way the participants *think* and *behave* toward their job and the productivity improvement challenge.'[134] The company's consultants were adamant that gainsharing ought not to be incorporated into the collective agreement. Such inclusion could render any program of gainsharing or continuous improvement subject to future erosion via the medium of collective bargaining.[135] Instead, union participation should be through a parallel structure of representation on a gainsharing team, while individual workers ought to be provided with the training that would allow them to understand the principles of gainsharing, as well as to take full advantage of it. In connection with this last point, Agrium also went on to introduce its competency development program, as discussed earlier.

A number of these points were objectionable to the local union, as well as contravening the international's guidelines. In response, the union drafted a letter of understanding on gainsharing that had three significant components. First, a clause on employment security was appended to simply state: 'Productivity improvements will not result in loss of employment for any employee.'[136] Unlike the earlier company proposal, which had placed a two-year moratorium on job severance, the union wanted more far reaching securities. Second, a provision was added that allowed for an independent auditing of all data pertaining to the plan. This independent audit was to be made available to the union on an annual basis. And finally, the draft letter contained a clause that would include any dispute over payouts within the regular grievance settlement process that existed between the company and the union.[137] This last component would have had the legal effect of bringing the plan under the auspices of the collective agreement, an outcome that the company was irrevocably opposed to. So, with this impasse, yet another letter of understanding between the union and the company went unratified.[138]

Significantly, however, the company has gone ahead unilaterally to introduce the gainshare plan, while providing assurances to its workers that lay-offs will not be the outcome of productivity improvements.[139] If anything, this furthers Agrium's engagement with a post-Fordist alternative where the role of gainsharing is to lend material support for the shift to a new culture of continuous improvement. Again though, the company may find itself in a position of paying for something that it does not sense it is receiving. Coming back to the original question of the future of continuous improvement at Agrium, it is probable that with the presence of a multitrained, non-hierarchical workforce and by the adoption of such initiatives as the gainsharing program, management will find itself under continuing pressure to proceed in a post-Fordist direction. These forces will also continue to be registered upon the union at Agrium and the decisions that it is confronted with in adopting a strategic direction.

Ultimately, what is significant is that workers at this facility have had a direct, ongoing experience with many of the most important aspects of post-Fordist work relations. Some of these features, such as multitask training and gainsharing, are now a permanent part of the employment relationship at the site, while others, such as the joint steering committee on continuous improvement, have been more transitory phenomena. However, this merely demonstrates that the transition into post-Fordism at established plants is often a complex, uneven, and indeed quite bumpy series of interrelated events. Agrium thus represents a more typical example of the real transition that is going on about us than an ideal construction might allow for.

Reprise

Four significantly different companies and cultures of employment at five different worksites have been described in Chapters 3 and 4. Table 4.7 presents a summary of the main highlights of the variation that was uncovered. It contains a schematic outline of the principal elements of diversity as they bear upon the discussion of Fordist and post-Fordist employment relations. It also provides a

TABLE 4.7
Summary of work organization at four firms

	PCS	Cameco	Central Canada	Agrium
Corporate mission statement	No	No	Yes	Yes
Joint steering committee/work teams	No	No	Yes	Yes
Job classification structures/CWS	Yes	Yes	Yes	No
Number of job classes	20	13	20	2
Collective bargaining format[a]	Adv	Adv	Adv	Adv/Joint
Profit/gainsharing	No	No	Yes	Yes
Lay-offs[b]	Freq	No	Infreq	No
Dominant managerial paradigm[c]	Loss Control	NS	EI/TQM	Cont Improv/TQM

[a]Adversarial; Joint problem solving
[b]Frequent; Infrequent; None
[c]Not specified; Employee involvement; Continuous improvement; Total quality management.

convenient reference marker for later discussions pertaining to the impacts that such paradigms have on those who have the most immediate experience of working within them.

A compelling case can be made for portraying these firms on a continuum that moves from traditional to post-Fordist work relations. PCS comes across as being the most traditional in terms of our operant criteria. The only slippage appears within the realm of lay-off behaviour. Even on this point, however, what is seen is not so much a deviation from Fordist norms, as an operationalization of those patterns under contemporary economic conditions. The use of a disposable, core workforce, such as can be witnessed at PCS, I have termed 'late Fordism.' With the exception of this point, there is not a great deal in Table 4.7 to distinguish PCS from Cameco. That Cameco has avoided a lay-off syndrome owes more to the different product market it serves than to any other factors. As well, there is no specified managerial paradigm at Cameco that holds the same place in human relations accounting, as one finds at PCS, with its devotion to Loss Control.

At Central Canada a mix of old and new is encountered. In terms of the job class structure, CWS, the corresponding wage grid, and the forms of bargaining that this gives rise to, there is little to distinguish Central Canada from the other sites, with the exception of Agrium. Coinciding uneasily with this structure, however, are several initiatives that have a definite post-Fordist hue to them. Employee involvement, with all of its accoutrements, has been entertained. Some, such as the special project, cross-functional work teams, are not currently active, although other initiatives, such as performance sharing, have become essential aspects of plant operation. This combination of features of traditional industrial relations and post-Fordist traits, and the contradictions between them, lends to Central Canada Potash the quality of a 'passive post-Fordism.'

Lastly Agrium appears as non-traditional on all of the criteria contained in Table 4.7. Specific managerial initiatives such as continuous improvement have been melded with a supporting structure of employment relations that is epitomized by the flat occupational/wage grid at this mine. Clearly, this firm has moved the furthest from conventional Fordist work relations.

Now the difficult questions begin. Do the variations that are highlighted in Table 4.7 actually make much difference in the labour processes or production politics that govern these workplaces? These are the issues of the next two chapters, commencing in Chapter 5, with an overview of the labour process and an analysis of the impact of Fordist and post-Fordist paradigms on the organization of work and the exercise of skill in the mining industry. A comparison of the skill and workforce influences of so-called industrial and post-industrial labour processes is also contained in this chapter. Chapter 6 studies Fordist and post-Fordist influences on the organization of the broader production politics in the workplace.

5

The Labour Process

The labour process has been eluded to on numerous occasions already with references to specific pieces of equipment (borers, bridges, and skips), working conditions (dust and fines), and procedures, such as compaction, flotation, crystallization, and reagents. It is now time to take the reader into a more detailed account of contemporary mine and milling techniques. Both the specific companies chosen for investigation in this study and the industry as a whole offer a rich vantage point from which to consider the contemporary labour process and claims made on behalf of post-Fordist and post-industrial theories of work.

In the first section of this chapter, the labour processes associated with the mining and milling of potash and uranium are described. The operant notion here is one of labour *processes*, for there are a number of them. Underground mining activity represents a highly mechanized pursuit, aiming towards full automation, but not there yet. Open-pit mining, which has become an increasingly common form of mining in such industries as coal, as well as uranium, is more akin to a heavy construction site. The study of both potash and uranium operations permits a consideration and comparison of both forms of mining activity. Meanwhile, modern milling industry is a prime example of a continuous process undertaking. As such, milling is not dissimilar to other refining and processing industries, which range from the petrochemicals sector to brewing. On top of this, each related but quite separate phase of extracting and processing raw materials has attached to it important maintenance functions. These jobs are enhanced by the capital intensive nature of modern mining and refining. Thus, apart from the various underground, open-pit, and mill operations, there is a sizeable maintenance workforce that retains claims to craft status. Skilled maintenance work is therefore very much a part of the labour process of mining work.

The diversity in production requirements and forms of organization found in the modern mining industry provides the context that is required for a rigorous examination of the claims of industrial sociology and labour process theory. Rather than constituting a relic, the industry as a whole presents a dramatic illustration of the effects of advanced technology and new organizational formats on a workforce that includes traces of traditional craft work, mechanized production processes, and continuous flow technologies. It is hoped that this will be of some interest to theorists of the labour process.

The second part of this chapter reviews the principal claims and debates that have swirled around industrial sociology and labour process theory. As we sit on the cusp of what many are considering to be a new industrial divide, the issues raised in these controversies are as relevant as ever. They may be organized around two different, if interrelated issues: post-Fordism and post-industrialism. A variety of interesting, but controversial claims have been staked out around each rubric.

In the third and last section of the chapter, this study's findings are analysed in the context of the theories of post-Fordist and post-industrial theses. This entails two different sorts of comparisons. First, there are the company by company comparisons, which allow us to control for industrial regime. Fordist and post-Fordist forms of organizing the labour process can be analysed through a comparison of the four firms that were introduced in preceding chapters. Then there are the post-industrial arguments that make certain suppositions about the changing nature of occupations and occupational requirements. With its interesting melange of line and flow operations and sophisticated maintenance work, the mining industry presents a fascinating case from which to examine the claims of post-industrial theory 'up close,' so to speak. In both instances, the effort revolves around grounding societal generalizations, whether they pertain to controversies related to post-Fordist arguments about Japanese production relations, or post-industrial claims concerning the new workforces, in a more detailed empirical analysis. In particular, the implications of new production relations for skill requirements is a special concern in the analysis that is to follow, owing to the importance that this issue has had for the labour process debate.

The Labour Processes

If the main challenge for the uranium operation that was studied lies in getting labour power to the site of production, for potash it is in getting to the ore body itself and extracting it from its natural environment. The geology of uranium mining, at least in its open-pit variant is relatively straightforward.[1] Uranium is found in veins and pit development follows them down through several benches (levels) that are roughly twenty-five to fifty feet apart. Mining consists of drilling, blasting, and hauling activities. The best analogy would be a large highway construction project with its massive pieces of earth moving equipment. At this point the labour process chiefly revolves around heavy equipment operation, where ore that has been blasted away from the main rock formation is loaded by hydraulic shovels into massive trucks to be scanned for richness, prior to being delivered to the crushing plant or deposited as waste material.

A typical pit crew at Key Lake consists of about eighteen workers directly engaged in mining activities. This complement consists of one driller, who operates a mobile drill platform, one or two blasters, a shovel operator, and approximately eight haul truck drivers. Rounding out the crew contingent are other equipment operators, including grader and scraper operators, backhoe operators, and the drivers of smaller vehicles. Unlike conventional forms of subsurface mining where different crews are simultaneously working out of different sections of the mine, in the open-pit operation at Key Lake, there is only one mining crew, working in tandem, per shift. Mining activity commences with drilling and the setting of charges to expose the ore body. However, instead of handheld or leg jack drills such as are used in underground hard rock mining, drilling in the pit is done by a large platform drill that is reminiscent of the type of technology used in oil exploration. Drill operation constitutes one job position, as does the setting of the charges, which requires the job holder to possess a blasting ticket. Loose ore is then loaded by the shovel operator into gigantic haul trucks of 50,000 kg capacity for a drive through the scanners prior to delivery to the stockpile.

The division of labour and consequently the occupational hierarchy in the pit is governed by the piece of machinery that the worker operates. Workers who operate the smaller pieces of equipment, such as the pump trucks that water the site down, the school buses that transport workers to and from the pit, and the front-end loaders that are used in clean up operations are at the bottom end of the job classification grid. As the equipment becomes larger and heavier, job classification and pay increase up through the five job classes that constitute the division of labour in the pit. Haul truck operators, the most numerous group of workers in the mining department, are responsible for taking the ore from the blasting site to the scanning and crushing areas and occupy an intermediate position in the pit. As Chapter 2 described, these positions are predominantly filled by Aboriginal workers. Shovel and crane operators occupy the top-end positions in the occupational pyramid of the pit, along with the drill operators and blasters.[2]

Production at Key Lake takes place on an around-the-clock basis. Over the course of one twenty-four-hour shift, the shovel operators provide continuous production, loading ore into the massive trucks that snake their way from the bottom of the pit up through the various switchbacks that connect the different benches to the stockpile area and adjacent crushing plant. There is nothing particularly exotic about much of this work. As mine operators, most workers labour in the relative isolation of equipment cabs. Haulage drivers described the boredom associated with making the same slow and tedious drive back and forth between pit and stockpile for twelve hours a day. Perhaps to alleviate such conditions the larger pieces of equipment now come equipped with air-conditioned cabs and FM music imported into the site from the west coast via satellite hookup. All operators are also connected to one another and to their supervisors by two-way radios. Instructions can easily be conveyed and accounts requested in what, in comparison with underground mining, is a highly visible labour process. Such visibility makes the tasks of monitoring and supervision quite straightforward. Unlike the subsurface work environment, shovel operators or haul truck operators would quickly be missed if they were to take any unscheduled breaks from the labour process.

In addition to the mining positions already discussed another three workers per shift are assigned to the 'de-watering' department, which is also positioned in the pit. This department is responsible for the operation of the pumps and pipeline that control the flow of ground water into the pit. The object of course, is to prevent ground water from seeping into the mining area. This entails the laying of pipeline and the operation and ongoing monitoring of the pumps that prevent pit flooding. Any water that does manage to enter the mining area is contaminated and must be piped to the mill for treatment in the bulk neutralization area. The de-watering works include the job positions of pump operator, pump helper, and labourer, which fall beneath the better paid of the heavy-equipment operating jobs.

Compared with uranium mining, the development of potash mining encountered more serious challenges. Chief amongst these is the nature of the geological environment through which mine development has to pass en route to the ore body. Potash deposits are found at comparatively deep levels. The mining operations that are included in this study are all undertaken at approximately one kilometre beneath surface level. Above the KCl deposits is a formation called the blairmore, a thick layer of loose shale and sand that is saturated with water under extremely high pressure. Sinking a shaft through this formation had to await the requisite developments in mining technology. This entailed inventing a method for literally freezing the blairmore prior to shaft installation, through the use of a circulating, refrigerated brine solution (see Figure 5.1).

This, combined with the use of bolted rings of cast iron tubbing to reinforce

FIGURE 5.1
Geological structure of potash mining

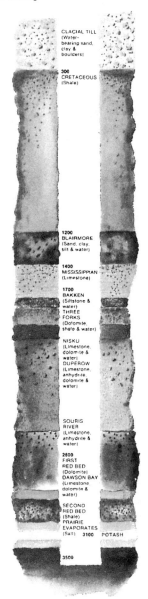

Source: Potash Corporation of Saskatchewan Inc.

the shafts, made it feasible to get through the blairmore zone to the potash deposits. Once exposed, ore bodies remain prone to flooding problems. Shortly after the Agrium mine was opened in 1969, for example, flooding problems caused a two-year shutdown. Other sites, not included in this study, have had to switch over to solution mining techniques because of water problems, or are plagued by ongoing water removal problems. Related to this, as potash and the surrounding geological bodies are soft rock formations, shifting ground is an additional complication that enters into the production process.

Potash mines that employ underground mining techniques deploy two shafts or head frames. This provides the basis for fresh air circulation in the underground environment. Personnel and materials are conveyed the one-kilometre distance in one shaft, while ore (muck) is skipped to the surface, on an ongoing basis, in the second shaft. Arrival at the bottom of the shaft takes less than two minutes and is akin to entering a well-lit underground parking garage. In fact, that is basically what the bottom shaft area is, a large, two-storey-high parking and storage bay for underground vehicles and equipment. The actual mining takes place at a considerable distance out from the shaft area.

Unlike the situation in open-pit uranium mining, there are a number of working faces in a potash mine, located at points of twenty kilometres or more from the bottom of the mine shaft. At the mines I visited, three or four mine faces were in production at any given time. Travel to the faces was through drifts 3.4 meters in height and 5.5 meters in breadth, or just wide enough to squeeze two Toyota '4 x 4s' past one another. Once outside of the service bays, the drifts are unlighted and not unlike travel on tertiary, gravel roads in the countryside, well after sunset. Travel time, once underground, may then amount to half an hour or so. On the way to a face, seldom are more than one or two workers encountered, if that many. This is truly a solitary environment.

This impression is reconfirmed at the working faces. All of the mines use what are known as continuous mining machines, or borers, normally with three-person crews. In other words, given the crew sizes, the number of mine faces being cut and the associated maintenance personnel, there is not a particularly large labour force underground, at any given time. At Agrium, for example, a typical underground shift would be composed of seventeen production workers parceled out over a tract of seventy-seven square kilometres.[3] Again the differences with the open-pit situation and the congregation of larger numbers of workers in one production area are stark.

The mining machines, or borers, are controlled by one operator, who is positioned atop the machine in an open cab. At the front of the miner are two three-armed rotors each fitted with tungsten carbide bits that chew out the mine face and crush the ore prior to automatically forwarding the muck onto conveyance equipment. Borer operation is a hot, dusty, and noisy job. On the cab temperatures hold in the 80–85 degrees Fahrenheit range, while visibility is limited to a few feet, owing to the dust conditions. Communication is possible only through the use of headsets, which is how the crew keep in touch with one another. For cutting, the line of direction is set forth by a laser beam, attached to the roof of the chamber and directed against the mine face. The operator must follow this line of direction in the operation of the borer, but retains a crucial control over the height at which the cuts are made. This is important as it ultimately controls the KCl content of the muck that is shipped to surface. Experienced operators can cut about one foot per minute, but it may take some years of experience to get up to this rate. According to some of our informants, the job requires a knack, or 'feel,' that some operators find very difficult to develop. With experience, the job may lose its challenge, but it does require an ongoing attention to cutting levels. Cuts that are made too high, for example, increase significantly the clay content of the muck that is forwarded to the mill, and consequently increase processing costs.

Two patterns of mine development are used in the industry and they determine to a certain extent what happens next. Increasingly common is what is known as the stress relief method of mining. This involves cutting 18-foot passages for approximately one mile into a drift and then backing out again with an additional 18-foot cut. Similar parallel cuts are made throughout the drift, with 24-foot walls left to separate the different cuts for support. In other words, the mine is developed as a series of long, straight, parallel works. This simplifies conveyance of the muck to the surface. An extendable and semi-permanent belt structure is attached directly to the miner and carries the muck to underground storage bins from where it is skipped automatically to the surface, as required for milling. Building the belt structure to transport the ore from the face to the skips is a significant aspect of underground work. At Agrium, for instance, there are some 32 kilometres of conveyor belt. Normally, three-person crews superintend the initial cuts into a drift. On the return cut back to the main passageway, the crew size may be dropped back to only two operators.

The second pattern that is commonly encountered is referred to as the chevron method. It involves carving rooms out at right angles to the main travel ways. Initially two passes of 81 feet are made, at which point a mobile conveyor belt is extended into the chamber and linked up with the miner. Other longer passes are then made at 45-degree angles to the left and right. This leaves a 50-foot pillar to support the room. Such chambers are cut out at 170-foot intervals off from the main tunnels. This technique of mining development requires the use of mobile conveyors, or bridges, that connect the miner to the permanent belt structure that runs through the length of the main travel ways. The bridge may be composed of several sections amounting to 300 feet or so of transportable belt line. The veyor, as it is called, is mounted on a series of rubber tires and follows the miner into the chambers that are being mined out. Associated with its operation are two back-up operators, one located at each end of the bridge. They are responsible for moving the veyor up to receive the muck as it comes off the miner and for keeping it correctly positioned to receive the feed. The back-up operators are also charged with maintaining an even flow of muck between the miner and the permanent belt structure, clearing away blockages and spills that may occur with oversized pieces of ore. This necessitates ongoing communication among the three-person crews which, given the distance separating them and the noise levels, can only be carried out with the use of two-way radios and headphones. To expedite this work further, veyors are now equipped with their own monitors and video cameras. This allows for automatic control and adjustment of the chute that funnels the muck from the miner onto the veyor belt.

If this were all there was to the labour process of underground potash recovery, it would indeed approximate a continuous process industry in its own right. But this is not the case. The crews are also responsible for the development work that precedes the actual mining. This includes securing the area that is being entered through roof-bolting and scaling activities, engaging in the work of belt extensions and belt repairs, and constructing the brattice which gets better quality air to the production site. During a shift, the borer operator must also perform periodic changes of the cutting bits on the rotary arms of the miner. Although this might seemingly provide some variety to the job, mine-face temperatures of 120 degrees Fahrenheit make this hot, uncomfortable work. A more accurate depiction of so-called continuous mining then would be a 'stop and go' activity. Aspects of it, namely, the actual cutting, do approximate a continuous process activity, but other dimensions entail hand labour or only slightly mechanized tasks. Notably, each crew is together responsible for the aforementioned tasks. In this respect then, and compared with other examples of soft-rock mining, the work has not been subdivided as far as it might have been.[4]

Despite what seem to be skeleton mining crews in today's industry, there are pressures to reduce and automate further. Management is of the opinion that only one back-up operator is necessary, now that feed rates are automatically monitored and controlled by video equipment located on the bridge. Union officials insist that with lengthy bridge extensions and in a noisy environment with poor visibility, this would constitute unsafe understaffing.[5] Remote control mining is also on the horizon. The potash companies have experimented with fully automated miners that can be controlled at some distance from the cutting face (in another part of the mine, or theoretically, on the surface). Although none of these prototypes is currently on line, it is only a matter of time. And if automated mining techniques do not make their first appearance in the potash industry, they most certainly will in the new high-grade uranium fields that are in the process of being opened up. Currently Cameco is developing new underground mine sites, where the richness of the ore body renders human exposure unsafe. This will absolutely necessitate the development of robotic miners.

The question of crew sizes and levels of automation ties in with a second issue, that of tonnage rates and production costs. Miners at each underground potash operation were aware of company production targets, whether they be daily tonnes of ore mined, or feet of rock-face cut.[6] But just as quickly, it was suggested that such objectives were guidelines only. The labour process of sub-surface mining was altogether too contingent to hold fast to strict production goals. In other words, no one took such targets too seriously. Whether they were obtained, or not, was no matter of great concern amongst the underground workforces. Interestingly, attempts that have been made to more rigorously monitor underground work have, to date, met with failure. At Agrium, for example, following the 1988 departure into scientific management, gauges were placed on the miners in order to measure the amounts of power that were being drawn. From this data, cutting rates could be estimated from afar. However, these devices were continually being broken and after a short while the company decided that they were more trouble than they were worth.[7]

Leaving the mine face, the labour process becomes even more highly automated. Skipping ore to the surface is now fully computerized in most of the mines. Bins, capable of holding 24 tonnes of ore are automatically filled and monitored by computer, before they are sent to surface every few minutes for delivery directly to the mill or for stockpiling in huge storage barns. Although this function used to employ a minimum of two workers, in most mines this work area is now completely deserted.[8]

Processing represents the second main labour process in mining activity. At this stage, the mineral is separated from the other material it is found with in the natural environment. With potash, the ore that is milled is of about 40 per cent purity. Processing involves extracting the KCl from the unwanted salts and clays that are part of the muck received on surface, manufacturing it to meet customer specifications, and disposing of the waste by-products. Each site included in this study has its own milling facilities. Uranium purity at the Key Lake site runs just under 2 per cent and, as a result, its recovery process is more involved than that of potash. Yellow cake is distilled from an ore body containing elements of molybdenum, nickel, arsenic, and radium that must be disposed of. Especially with open-pit uranium mining, the milling operations appear to be the defining moment and most challenging technical aspect of the labour process. This distinguishes the industry from potash, where the organization of labour processes in a voluminous underground environment presents the main technical challenge.

Typically, from the outside, a mill, whether it be potash or uranium, resembles a multi-storeyed, medium-sized, windowless factory. The first step in the refining process entails transforming the ore into a medium that can be worked upon. Potash is conveyed by belt to the mill complex where it is delivered to a number of impactors to be crushed and screened into granules. This process is totally automated. Workers are only brought into this part of the mill for occasional clean-up activity.

Following impaction, the ore is converted into a slurry through the use of a brine solution in a process known as scrubbing. The brine dissolves the clay, leaving the potash crystals intact. The solution passes on to the floatation vats, where any remaining clay particles are removed. Sodium chloride is then separated from the potash in the reagents section, where guar, amine, flotation oil, and frother are mixed with the slurry and agitated. The potash particles attach themselves to the resulting air bubbles and are skimmed off the top of the solution with sets of continuously rotating paddles. The wastes sink to the bottom of the reagent vessels and are siphoned off as tailings. Any remaining brine is removed in an automated centrifuge.

At this point, the process of recomposition commences. The potash solution is passed through gas-fired dryers, where heated air is passed up through the KCl crystals in a process that is not dissimilar to the making of popcorn with a hot air convection popcorn maker. In this instance though, we are left with dried KCl crystals which are then screened into products of various sizes and qualities, and which commonly go under the generic names of 'coarse,' 'standard,' and 'special standard.' As at the start of the process, the screening that takes place here is fully automated, with no direct worker involvement.

Additional value-added product may be created at this stage through processes of compaction and recrystallization. With compaction, undersized potash granules are placed under intense pressure and heat in compactors to create manufactured chips of potash. These are recrushed and screened into marketable course or granular products. Additional reclamation may also be undertaken through the collection of fine dusts. Sophisticated dust collection technologies are utilized during the drying, compacting, and screening processes. This output is then dissolved and reformed into potash crystals in large, multi-storeyed recrystallization towers. Both compaction and recrystallization procedures create a comparatively high value-added final product. Granular potash is a high-demand item in the mature North American corn, soybean, and lentil markets. White product is the principal output of the recrystallization process.

The milling process is technologically sophisticated and capital intensive in the extreme. The processes just described would be staffed with no more than eight workers. These numbers are reflective of ongoing technological downsizing. At Central Canada Potash, for example, in the early 1980s, thirteen operators were employed in the mill. Currently, through a combination of control room centralization and job combination, six are employed in the off (night) shift, including two control room operators and four floor workers.[9] At Central Canada and the other potash sites, the operation is run out of a central control room located in the mill, although the centralization of these functions is by no means a technological necessity. The control room is filled with circuit monitors that relay information to the control room operator(s), who is always in attendance. Critical values, such as the levels in the tanks, and the overall level of throughput in the mill are monitored from this point, in a closed loop feedback system. The control room operator is also responsible for sending out specific instructions to the floor operators via two-way radio. Each floor operator, in turn, is in charge of one or more areas such as flotation or reagents. Everyday activities would include sampling the product, dealing with specific problems, such as spills, that cannot be dealt with from the control room, and finally overseeing the area for smooth operation. The working environment on the floor is noisy, warm, and sometimes smelly. What perhaps is most striking to the observer, however, is the dearth of labour power in the modern mill. One could easily do a tour of such a workplace without encountering any workers on the floor. Contemporary milling can therefore be considered an automated, continuous flow, 'post-industrial' process, par excellence.

From the mill, the product is conveyed to a huge indoor storage barn, which literally contains mountains of stockpiled potash. This is drawn on as required by the loadout department, which is nominally responsible for the loading and securing of the railway cars that carry product to markets in the United States or to Canadian port facilities. The loadout department at each mine will normally employ around five operators on one shift, although during the busiest shipping seasons, a second shift may be added. Tasks associated with this function are performed in tandem from a control room, where belts and shoots are operated by workers who are responsible for opening and securing the hatches on the specially designed railway cars.

The organization of work at the Key Lake uranium mill offers some interesting variations on that of the potash milling process. Preparation of the ore for refining takes place in a separate grinding mill, where stockpiled ore is fed by front-end loaders down through a 'grizzly' screen and into the plant via a feeder conveyor. Here, the ore goes through two separate grinding phases, where it is first reduced and mixed with water to produce a slurry, and then is further reduced and screened in a separate section (the ball mill) to produce a fine, wet, sandlike material. The grinding mill is staffed with two operators, one in the control room and one out on the floor. Once the mill has been fed with raw material, the actual grinding process has been fully automated.

From this operation, the uranium slurry is piped a distance of 1.5 kilometres to an adjacent mill facility where it is held in large storage pachucas. Each one of four vessels holds between six and eight hours' worth of production. From the holding pachucas the slurry is fed into the leaching area where it undergoes two treatments (see Figure 5.2). First, in primary leaching, slurry is mixed with the acidic output of a process located further downstream. It then flows by gravity to the primary thickener, where the solids are fed into the secondary leaching circuit and the solution on the top is pumped to the solvent extraction plant as 'pregnant solution.' In the secondary leaching process the ore solids from the primary leaching are combined with sulphuric acid and oxygen at elevated temperatures and pressures. This circuit employs three operators.

After spending about five hours in the leaching circuit, the slurry passes to the washing area. Here, the solids that are dissolved out of the leaching process flow through an eight-stage thickener circuit. This process separates the uranium solution from waste solids that are neutralized with lime and sent to the tailings storage area. The uranium solution, meanwhile is passed back to the leach area, where it is fed as 'pregnant solution' to the solvent extraction unit. At this stage the uranium is extracted from the solution by mixing it with water and kerosine. The output of the multi-staged solvent extraction process is a purified uranium solution resembling orange juice. As in the other areas, the solvent extraction unit is staffed with two operators, one remaining in the unit control room and the other monitoring the floor area.

The loaded strip solution, as it is referred to at this point, is then forwarded on to the yellowcake precipitation zone. The uranium is recovered by adding ammonia to the purified uranium solution and feeding the product into an agitation tank. This causes the uranium to precipitate as ammonium diuranate, which is subsequently forwarded to a thickener and a centrifuge. The solids that are produced in the centrifuge are sent on to a multihearth furnace, where the ammonium diuranate is dried as yellowcake. Despite the name, the final product is a blackish-green powder. One to two operators are responsible for packing it in 200-litre steel drums and loading it onto covered flatbed trucks for shipment to Ontario or the United States.

Separate sections of the mill remove the ammonia from the purified uranium solution as ammonium sulphate crystals, which are processed into fertilizer, and

FIGURE 5.2
Key Lake Mill simplified flowsheet

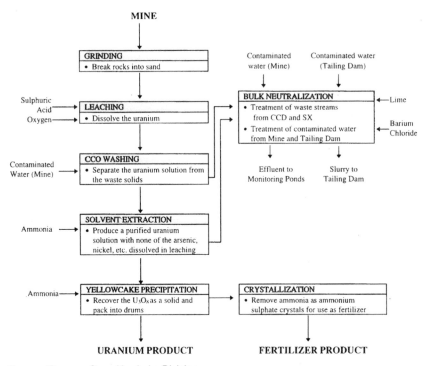

Source: Cameco Corp. Key Lake Division

treat the waste product. In the latter process, called bulk neutralization, waste from the solvent extraction process and contaminated water are treated for removal of acids, radium, arsenic, and nickel. Solid waste products are deposited in mined-out pits. Liquid wastes are treated, deposited into a monitoring pond, and then released into the environment. Bulk neutralization, as with the other areas, is customarily staffed with two operators.

Uranium milling employs more workers than potash milling. As opposed to the six to eight operatives customarily found in a potash mill, an average of twelve workers are employed on each shift in the mill at Key Lake, plus two operators in the crushing facility.[10] Coinciding with the larger numbers is a somewhat greater dispersal of control over the technology at the Key Lake mill. Instead of centralizing control of operations as is the standard in the potash industry, each circuit of uranium refining has its own control apparatus. This makes for a total of four control rooms, each with its own operator, plus an additional floor operator assigned to the area. Four 'floating' operators round out mill staffing.

Adoption of this control structure at the Key Lake mill was both optional and experimental. The mill had been originally arranged in this fashion until mill functions were reorganized and brought together in one centralized control room.[11] This allowed for a savings on labour power, although no direct lay-offs occurred as a result of this move. Former circuit operators were placed on a newly formed 'day crew,' which was responsible for housekeeping functions in the mill. This crew also functioned to provide a labour supply buffer against employees who missed the plane into camp, or otherwise did not show up for work. Apparently this reorganization was not well received by those employees who were reassigned to the day crew. The reassignments were associated with a loss of control over the use of the technology in the refining process and consequent perceptions of deskilling. Palpable job dissatisfaction, as well as a deterioration of physical conditions in the mill resulted. Consultants were commissioned to look into the problems and following this, management acted on their recommendations to return to the original system of multiple control rooms exercising control over specific facets of the refining process.

Do such organizational choices matter, in any significant fashion, when it comes to job demands and shifting trends in occupational requirements? In the next section the debates around work organization, technology, and trends in employment profiles are revisited and updated. Then some of this study's evidence from the multiple labour processes associated with modern mining industry is brought to bear on these important issues.

Labour Process Theory

Three previous literatures have a direct bearing upon the labour processes presented in the last section. The first is most accurately located within an earlier industrial sociology tradition and has more recently been revived in post-industrial discourses. Without denying the importance of Taylorism or Fordism, these organizational formats are portrayed as passing stages of industrial evolution.[12] As proffered by one spokesperson, 'Once we understand Taylorism as a process conditioned by historical developments, we can imagine that it may itself be superseded by other and different approaches to work, not because workers or capitalists will it, but rather because new developments in technology or market structure modify the motives and actions of managers and workers alike ... postindustrial technology makes both Taylorism and its critique increasingly irrelevant.'[13] Accordingly, the last third of the twentieth century has been marked by an exit out of Taylorism, brought about by developments in modern technology, specifically the automation of an increasing number of tasks and their reduction to continuous flow types of processes, such as those described in the last section.

From a second perspective, Taylorism and its accompanying forms of technology and industrial relations represented an irreversible dynamic; a capitalism made fully conscious of itself, a totally discursive managerial practice.[14] Far from losing relevancy, the scientific management movement was destined to take over more and more fields of employment organization. In the process, technology, including its most contemporary guises, rather than constituting an independent variable, has been fully shaped to correspond to the principles of scientific management.[15]

Finally, a third approach views the issue of work organization as a matter of strategic choice. Various formats for organizing the work-effort bargain are potentially available. There is no inexorable logic either leading into or out of Taylorism. Instead, managers and workers have available to them a host of alternatives. Which one is selected will depend upon a number of factors, including the existent state of relations between workers and employers at any given point of time.[16]

Thus three different images of organizing labour under capitalism exist: a series of evolutionary stages into and out of industrialism; a representation of a deeper underlying logic of accumulation that structures work organization; or a matter of strategic choice.

For a previous generation of industrial sociologists, the coming of computerization and the possibilities for fully automating work processes were often portrayed as providing a respite from the problems of industrial alienation and conflict that characterized much industrial organization. Joan Woodward saw parallels between the work situation presented by contemporary continuous process industries and small batch forms of craft production, in the first of many analyses that have suggested a return to the conditions of craft labour.[17] Both continuous process technologies and craft production were characterized by their use of small, more cohesive working groups and high ratios of skilled to unskilled labour. Moreover, in addition to the high ratios of skilled maintenance personnel in the continuous process environment, the nature of the work itself tended towards an upgrading of the operator functions. For Woodward this type of technology removed many of the manual aspects associated with skill, while retaining the cognitive functions associated with goal attainment. As she went on to explain, employees in the newly automated labour procedures of continuous process industries 'although often highly skilled were not formally recognized as skilled outside their own firm. The traditional differentiation between the skilled and the semi-skilled worker does not allow for a situation in which the manual and motor elements of skill have been taken out of the main production task, while the conceptual and perceptual elements remain.'[18]

Although continuous process technologies originated in the manufacture of specialized liquids and gases, Woodward suggested that they were increasingly coming to characterize production processes in such well-known pursuits as steel, paper, and food manufacturing, to which mining and milling activity might also be added. Associated with the adoption of this technology was the employment of fewer numbers of better-paid workers, with increased skill mixes and some skill enhancement for the operating class. These conditions, it was assumed, would lead to an overall improvement in the industrial relations climate. Process industries were also, according to Woodward, conducive to longer lines of managerial command, smaller spans of control, and a growing ratio of managerial and supervisory staff. Intriguingly, she did not consider that the latter traits might in fact mitigate against the greater levels of autonomy and skill that were also associated with this industrial form.

At about the same time, Robert Blauner compared four American industries and reached very similar conclusions. The loci of his research were a printing shop, a textile mill, an automotive assembly factory, and a chemical plant.[19] For Blauner these sites epitomized conditions that were characteristic of craft, paternalistic, mass production, and continuous process industries, respectively. Craft production minimized all aspects of alienation. Workers controlled entry into the trade and the pace of work. Lengthy apprenticeships and the mysteries of the trade acquired therein produced craft identities and substantial communities of interest. As Blauner saw it, conditions in continuous process industries replicated many of the conditions found in craft workshops, including high levels of job security, and a high-skill mix among the workforce. Even for the operatives, Blauner noted: 'In the place of physical effort and skill in the traditional manual sense, the major job requirement for production workers in continuous process technology is responsibility.'[20]

Indeed, this was to be the hallmark characteristic of such industries. As compared with the tyranny of the assembly line, or the authoritarianism of the textile mill, continuous process technologies offered workers enhanced levels of responsibility that were capable of lending work new meaning, as well as extended opportunities for self-fulfilment.[21] Generally such jobs encompassed a wider range of operations for individual workers than was customary in alternative technological environments. Still Blauner was aware of certain ambiguities that he chose not to analyse further, as witnessed in the observation that 'Continuous process technology ... contains tendencies toward greater interest and involvement and toward greater monotony and boredom.'[22] He acknowledged that 'the worker's role changes from providing skills to accepting responsibility.'[23] Nevertheless, this 'reverses the historic trend toward greater division of labour and specialization.'[24] Like Woodward, Blauner concluded that the emerging technical conditions in modern industry were at the very least impelling a relative skills revision among those workers who were using the new processes.

Subsequently these same themes have been expanded upon under the rubric of post-industrialism, which has lent a further degree of specificity to the notions of work and skill in computer-controlled environments. In a fully automated process, there is precious little left for the remaining workers to execute. Ergo, according to Hirschhorn, the very distinction upon which Taylorism was found, the division between conceptualization and execution, is passé.[25] So too is the other operant principle of scientific management, namely, the perfection of motion through the elimination of error. Instead, in automated feedback systems such as are found in uranium and potash mills, error is imported into the production process, to be overcome in internal self-correcting iterations. Cybernetic systems of production are thus continually adjusting to anticipated changes in the environment, or so-called first-level errors. In highly complex, interdependent systems, however, not all errors or sources of variation can be anticipated. In an industry such as potash, for example, the quality of the ore that is being skipped to the mill may vary in unexpected ways, depending upon the nature of the seam that is being mined and the skill of the borer operator. To deal with this, mill operators must develop new sets of diagnostic skills in which automated systems are manipulated to achieve optimal levels of added value.

This is said to reverse the deskilling trends associated with Taylorism. New skills, founded upon density of perception, breadth of vision, and depth of theoretical knowledge, are brought into being with the advent of automated labour processes. In the wake, skills are extended by what Zuboff refers to as the unique informating capacities of the new technologies, as opposed to being replaced by them.[26] Workers become problem solvers, engaged in the ongoing developmental work of continually adapting the flexible new technologies to the creation of superior quality and new products. Historically then, the worker moves 'from being the controlled element in the production process to operating the controls to controlling the controls.'[27] Employing more complex technology, and called upon to sort out more difficult, second-level errors, 'Successive waves of automation can thus be expected to reduce workers' low skill workload, leaving them with a higher overall skill level.'[28]

Zuboff has given the name of intellective skills to the new job requirements.

Relying on a comparative study of different employments, including three pulp and paper mills, she argues that a radical distinction exists between the skills that were called upon in the exercise of craft labour and skills that are used in an informating environment, in a fashion that is partially reminiscent of Joan Woodward's earlier work. Traditional craft skill is premised upon forms of non-discursive, sentient knowledge that is acquired by the body through labouring experience. It translates into abilities to 'act on.'[29] More commonly, such skills are associated with practical know-how, tacit knowledge, and the tricks of the trade. Intellective skill, on the other hand, is associated with the ability to relate electronic text to the real physical labour process and to make use of such symbols to manipulate the production process in desired ways. Computer-generated text now interpolates the relationship between 'work' and 'action.' In short, mental labour increasingly replaces manual labour for the operatives in an informated labour process.[30] According to some, the existence of new skill sets are indeed validated by wage trends and the failure to realize significant wage savings when craftwork, such as metal machining, is automated through the use of numeric control technologies.[31]

These arguments are salient for the computerized potash and uranium mills that form part of this study, as well as for the highly mechanized processes that are undertaken in continuous mining technologies. Later in the chapter, the empirical evidence collected from these operations that pertains to the theses of the post-industrialists will be examined. At this point, I simply wish to draw attention to one theoretical ambiguity in much of this genre. There is a persistent element of technological determinism to post-industrial analyses of the labour process. Developments in technology, and in particular the diffusion of cheap sensing devices and the corresponding arrival of more flexible machine forms, has rendered Taylorism and Fordism obsolete. This has made Taylor's vision of human labour power redundant, if not counterproductive. Interpretive, diagnostic, synthetic, and social skills replace the deskilled, fragmented worker of yesteryear. However, this vision remains difficult to reconcile with the more normative prescriptions that the post-industrialists usually attach to their arguments. Here analysis gives way to advocacy, and managers are urged to transcend Tayloristic practices for integrative, learning environments. Once again, as Hirschhorn suggests, 'we must design jobs in such a way that workers can effectively control the controls ... To do so, we must transcend our Taylorist inheritance and develop a new theory and practice of job design.'[32] Unconsciously, and uncritically, the argument has slipped over into one of choice.[33] Although the new technologies may be more compatible with holistic, team-based approaches to work organization, managers may not operate on this 'knowledge.' Taylorism may apparently still be in need of transcending!

It remained for Daniel Bell to take these arguments one step further, into the realm of societal trends. For Bell, changing occupational structures and the associated skill levels that accompanied them were responsible for the shift into a post-industrial society. This transition was synonymous with the growth of services and especially a public service sector, the expansion of sociologizing modes of calculation, and the explosion of a new class of knowledge workers, responsible for staffing the institutions of the post-industrial world.[34] Parallel with these developments was the erosion of the semi-skilled working class of industrial society. As Bell projected, without a hint of critical acumen, semi-skilled employment will shrink until, by 1980, it will rank in third place, 'outpaced by clerical [labour], which will be the largest, and by professional and technical workers.'[35]

Since Bell's foray into post-industrialism, other authors have updated and sharpened the argument. Clement and Myles project upgrading effects associated with more flexible forms of accumulation and a post-industrial occupational structure. For these authors and others, the shift to a service-oriented economy appears to be the most important factor in swinging the balance in favour of creating new more highly skilled workforces. Thus according to Clement and Myles, 'Although much less than a knowledge revolution, the net result of the shift to services has been to increase the requirements for people to think on the job.'[36] But parallel to this runs another argument wherein changes in the mode of organizing labour processes accounts for an upgrading of working skills across social classes. It is unclear, however, whether skill enhancement is principally a product of *what* is being produced – services instead of goods – or *how* they are being produced. Although we do not examine changes in the occupational division of labour in this book, the labour process needs to be considered in its own right, and here it is not enough to equate an abandonment of Tayloristic philosophies with job upgrading. This connection remains to be investigated in Clement and Myles' research.

In a similar vein, and employing a large sample of British workers, Gaullie argues that workers who have remained in the same job are most likely to report either skill enhancement, or skill stability, depending upon their specific location in the occupational structure.[37] Deskilling, on the other hand, was an infrequently reported occurrence. These trends were also magnified when the use of automated technologies were brought into play. Regardless of occupation, the use of such technology was associated with longer training periods, greater levels of in-house training, and longer familiarization periods, which in turn are all taken to be indicators of skill accretion. Although we may wish to question the uncritical use of some of these indicators as proxies for skill, post-industrial theory, in both its early and contemporary guises, has remained consistent in arguing against deskilling trends.[38]

It was against just these types of conclusions, at least in their initial iterations, that critical labour process theory was directed. This body of research commences by inverting the order of causality that had informed existing industrial sociology. In place of the technological determinism that had accompanied analyses of continuous process industries, and despite differences of emphasis, Braverman, Noble, Shaiken, and others asserted the priority of the social purposes of work.[39] Both theoretically and historically, technological advance plays second fiddle to the social organization of employment. In turn, the organization of labour could not be divorced from the objectives of production, and post-industrialism notwithstanding, this has remained the valorization of capital through the suppression of production and circulation times. Later theories of scientific management simply brought this internal dynamic to the level of managerial discursiveness.

Given the overarching purpose to which labour is organized in capitalist economies, the fragmentation of work into specific tasks that are matched with corresponding skill and pay gradients is *the* constant, unalterable feature of the capitalist labour process. First identified by Charles Babbage, Braverman goes on to observe: 'Babbage's principle eventually becomes the underlying force governing all forms of work in capitalist society no matter in what setting or at what hierarchical level.'[40] Economizing on the use of labour power thereby becomes a sine qua non of successful accumulation. This carries with it two tendencies. First, for any given task, the skill of the worker and corresponding rate of pay should be strictly calibrated with the requirements of the jobs that need completion. Less expensive labour is, obviously, preferable to more expensive. There is no point in paying for skills that are not required. Second, and following from this, the labour process should be organized so as to reduce the wage bill through the substitution of unskilled labour for skilled. Systematic job observation, managerial planning, organization of the work-effort bargain through the differential piece rate, and the modification of technology to incorporate skill and discretion in machinery that is controlled by management, all become elements of scientific management. Each is designed to further the division of labour and thereby increase the efficiency with which labour is exercised, as well as the attractiveness of the work-effort bargain.

For Braverman, the counterpart to this logic is the progressive elimination of the skilled craftworker and of the control exercised by this labour force over the conditions of production. Jobs of lesser skill and diminished rates of pay are substituted for holistic work practices. The process leads, inexorably, to continuous job fragmentation, the deskilling of labour, and further alienation, in spite of the human relations initiatives that are periodically called upon by corporate management to resolve problems of job dissatisfaction and employee apathy.

Throughout the analysis, the project is seen to be as insidious as it is successful. 'The human instruments are adapted to the machinery of production according to specifications that resemble nothing so much as machine-capacity specifications. Just as the engineer knows the lubrication requirements, etc. of a motor according to a manufacturer's specification sheet, he tries to know the motions of a given variety of human operator from standard data. In the system as a whole, little is left to chance, just as in a machine the motion of the components is rigidly governed; results are precalculated before the system has been set in motion.'[41]

As implied, scientific management, or Taylorism, renders conscious the full logic of capital accumulation. It becomes *the* discursive practice of modern management, while those who are subjected to it bear the full weight of its effects in loss of skill and job degradation. These tendencies show no regard for economic sector or occupational status. Although originating in industry, Braverman makes it amply clear that the principles of scientific management are equally extendable to the burgeoning service sector and are compatible with a range of technologies, including continuous process industries.[42] Even the lower to middle echelons of management are themselves subject to the rationalizing effects of greater job partition and downward pressure.[43]

In comparing the diagnoses arrived at by Braverman with the conflicting trends that are depicted in industrial sociology and post-industrial analysis, two methodological points are worth noting. The first is the question of benchmarks. Quite clearly, differing bases are sometimes used from which to draw conclusions about contemporary directions. For the industrial sociological literature cited above, the implicit benchmark is mass production, assembly line industry. Making comparisons with this particular organization of work, claims are set forth concerning continuous process technologies and the enhanced skill requirements that they allegedly exhibit. Much of the post-industrial analysis has been based on similar foundations, although some have argued that the new intellective skills demanded by flexible technologies are simply incommensurate with traditional craft skills.[44] In other words, new skill demands are different from, but, at the very least, equal to traditional craft skills. Such claims, however, still lie largely within the realm of assertion rather than demonstrated fact. Braverman, on the other hand, adopts a longer-term, historical frame of reference. Using the earlier dimensions of craftwork as his benchmark, he finds that the skill demands of contemporary employment pale, irrespective of technological mix. From this longer-term perspective, even the analytical categories that are today employed as part of the industrial census, such as the tripartite breakdowns into skilled, semi-skilled, and unskilled work, are no more than artefacts that have assumed the form of powerful reifications.[45] Thus, benchmarks are key to this discussion. The advantages of adopting a long-term historical view in the study of social trends are self-evident. At the same time, however, these advantages may not always be the most relevant. Workers tend to compare their situation today with what it was five or so years ago rather than with the experiences that were encountered by their parents and grandparents. Braverman can and does ignore this point, which raises a second metamethodological point.

At the outset of his work, Braverman acknowledges the distinction between the objective and subjective elements in the theory of social class. *Labor and Monopoly Capital* is reserved largely for the former – the long-term transition of the capitalist labour process towards deskilled and degraded forms of employment.[46] For certain purposes this distinction is not only unavoidable but also highly helpful. In other respects, we need to be more wary. As Burawoy notes, a complete theory of the capitalist labour process, which includes analysis of control, resistance, and hegemony, must of necessity consider the experience of work and perceptions of that experience.[47] To that I would add that our very notions of skill itself have an inherently social and hence subjective dimension to them. Or, as several feminist authors have argued, skills are socially constructed to the extent that they are validated by a public as real skill.[48]

Consider, for example, the influential research of James Bright, upon whom Braverman builds much of his evidence for the deskilling thesis. Bright utilizes two indexes, one for different levels of mechanization and automation, and the other, a composite index to gauge skill requirements for varying degrees of mechanization. In effect the latter is a job evaluation instrument almost identical to the CWS scheme discussed in Chapter 3 and in use at three of the worksites included in this study. The composite index comprises, amongst other items, estimates of 'general skill,' defined as the 'understanding and ability in a task'; mental effort, including abilities to 'sense and analyze job requirements,' and 'attention and concentration'; and 'unpleasant environment conditions or work arrangements.'[49] Bright allows that in many instances these indicators possess 'a highly subjective connotation,' and I would concur.[50] The subjectivity of skill measurement is one of the reasons why job evaluation *committees,* replete with recourse to full arbitrational mechanisms, are a necessity.

In short, skill is a multifaced, evaluative notion. It is at once an objective attribute, a social construction, and a political resource for those who can stake out claims, perceived as legitimate, to its possession. As Bright suggests, for any given dimension of skill, deskilling may occur at different levels of mechanization and on different pieces of equipment. Alternatively, Bright argues that deskilling may be offset by counteracting trends such as job enlargement or skill broadening in certain trades. The development of electronic circuitry is often taken as an example of these latter trends.[51]

With respect to the industry under consideration in this text, previous findings have also been divided. Researchers in Canada and the United States have by and large viewed the development of the mining industry as providing confirmation of the trends pinpointed by Braverman, in the form of greater task subdivision, and a general deskilling of labour. Hence, 'The early coal miner was ... an independent craftsman who worked largely without supervision, and mining was a craft occupation with counterparts in other industries of the nineteenth century.'[52] The subdivision of tasks, as in the coal industry, into full-time timberers, shot-firers, track layers, and coal getters, as well as the introduction of mechanized under-cutters, is viewed as having made some inroads into the skilled position occupied by miners.[53] Mechanization is portrayed as having, at first, contradictory effects on skill levels, and then, ultimately deskilling consequences.[54] In evidence that pertains directly to potash recovery, the introduction of continuous mining machines in the coal industry 'has necessitated new dexterities and technical knowledge.'[55] This has been offset, however, first by more intrusive managerial specifications as to how the work is to be done along with greater task subdivision, and latterly by fixing more elements of the control of the mechanized long-wall miners in the machines themselves.[56] As in potash, the coal operator controls the height at which the machine cuts, but not the fixed path along the face that is being cut out.

Clement reached similar conclusions in his exhaustive study of hardrock mining at the nickel giant, Inco Ltd.[57] According to Clement, owing to changes in mining technique and the organization of work, 'Management now has more direct control over the work process, workers themselves are being ordered in a hierarchical structure, and there is a clear tendency toward the loss of craftsmanship in the art of mining.'[58] Similar consequences are observed in the milling labour processes, where much of the former craft production has been replaced by 'instances of dial watching and patrol duty.'[59] For Clement, and contrary to the critics of Braverman, what is being witnessed is a movement away from systems of responsible autonomy, 'so that miners are becoming machine tenders or machine operators ... subject to a detailed division of labour' and closer supervision, while mill operatives are becoming mere patrol functionaries. This creates the conditions 'in which unskilled labour can be brought into the organization and quickly trained through standardized methods to operate or maintain a number of specified pieces of equipment, thus reducing the need for experienced or skilled workers.'[60]

In his analysis Clement alludes to other managerial options, but comes down squarely on the side of Braverman in viewing scientific management as the operant motif of the contemporary mining industry.[61] This view is seconded by Alan Hall in later research that was undertaken at the same Inco field sites. Once again, Hall argues that in the transition from conventional to mechanized to bulk (that is, continuous) mining techniques, there has been a growth in job specialization and a consequent reduction in required skills.[62] But Hall also recognizes that labour force downsizing in the mines has brought about an expansion in job duties, such as miners taking on both support and maintenance responsibilities.[63] These two trends remain theoretically unreconciled in his analysis. Moreover, as detailed in Chapters 3 and 4, scientific management is only one option among others for organizing mining labour.

Intriguingly, European studies of the industry have come to rather different conclusions. Comparative studies of conventional longwall and composite longwall systems in the British coal industry have demonstrated the dysfunctionality of a rigid division of labour that is associated with the former organization.[64] With the longwall method, each of three shifts is assigned a different specialized function; cutting, filling, or advancing. According to Trist and his associates this method of organization was largely unworkable. It fostered high rates of absenteeism, calculated forms of irresponsibility and indifference, and endless bouts of minor pay negotiations for completing work that should have been done by previous shifts. In itself, Trist understands mining to be a highly skilled occupation that entails both a preparatory cycle and a production cycle. In separating these tasks, management is depriving itself of the composite work group skills that self-selecting teams of miners (marrow groups) brought to their work.[65] By switching back to team-oriented production (composite systems), management could recapture these gains in a mechanized work environment.[66]

Organization in work teams that both internally assign and rotate jobs would in some cases allow for multiskilling, or at the very least, skill retention, even as mechanization proceeded.[67] Ultimately, Trist saw the replacement of conventional longwall by composite forms of organization as leading to better quality work, less supervision, greater workforce flexibility, and the replacement of job alienation with a task-oriented commitment. Comparisons that he had already conducted between sites using the two forms of organization demonstrated reduced rates of absenteeism and higher levels of productivity where responsible autonomy had been reinstituted.

For Trist and his associates then, the mode of coal extraction was ultimately a matter of organizational choice. The strictures of scientific management *could* be subscribed to, but they were not necessarily optimal. Alternatively, other forms of organization, such as systems which stressed the autonomy of work groups, could be instituted. Technology, although a constraint, was not a uniquely determining factor in the overall organizational picture of coal mining or other labour processes.

Trist's work on the British coal-mining industry has a strangely contemporary ring to it. Some years later, Friedman would generalize Trist's findings to argue for the omnipresence of two broad managerial control strategies that he termed direct control and responsible autonomy. Direct control strategies are synonymous with Taylorism and its job fragmentation, obtrusive supervision, and reduction in individual responsibility. The responsible autonomy option, on the other hand, 'attempts to harness the adaptability of labour power by giving workers leeway and by encouraging them to adapt to changing situations in a manner beneficial to the firm.'[68] In other words, responsible autonomy treats labour power as variable and attempts to harness it to the interests of capital, rather than to suppress it.[69]

For Friedman and others, adoption of a particular form of managerial control represents, in essence, a strategic choice. Most notably, the choice is influenced by the levels of resistance that management encounters, as well as by the financial position of the firm. Militant workers, possessing needed skills and working for financially healthy companies or in thriving regions of a national political economy, are more likely to be confronted with strategies premised upon responsible autonomy options than are others.[70] As working conditions under a regime of responsible autonomy are viewed as being superior to the alternative of direct control, both in the earlier work of Trist and by Friedman, such workers are treated as belonging to a core group of better-off employees. Correspondingly, workers who find themselves in a weaker position in skill, job security, employer prosperity, or regional gradients are more likely to encounter strategies of direct control or Taylorism in the workplace. Because these strategies are perceived as inferior such workers are placed within a labour market periphery.

On this last point, that is, on the factors that are likely to produce a strategy of responsible autonomy as opposed to the adoption of Tayloristic work practices, Friedman's analysis appears to be the most dubious. In North America at least, one of the main manifestations of responsible autonomy is to be found in programs such as Quality of Working Life ventures, which place a renewed emphasis on teamwork and responsibility. Such initiatives, however, have been introduced precisely in the wake of declining levels of worker militancy and growing employment insecurity, just the opposite of the conditions that Friedman views as propitious for pursuing responsible autonomy formats. Recognition of this point leads immediately to a reconsideration of the claims that have been made on behalf of such systems, at least as they are often encountered in Canada, including those arguments that pertain to skill enhancement and political empowerment in the workplace. This in turn, leads into the realm of post-Fordist discourses.

Questioning the conditions under which current strategic choices are made serves to recast the debates over direct control and responsible autonomy in a new, more contemporary light. For the most part, theorists of responsible autonomy have remained silent on the question of post-Fordist work relations and especially those elements associated with so-called Japanese production relations. This is unfortunate. One exception is Tom Rankin's work on new forms of work organization, which explicitly adopts the socio-technical systems approach of Trist. Rankin identifies similarities between Japanese production relations and the systems of responsible autonomy originally proposed by Trist, although for Rankin the socio-technical approach places more emphasis on adapting technology to users than is found in Japanese plants.[71] That aside, and although a direct lineage between socio-technical notions of responsible autonomy and post-Fordist production relations is clearly not the case, there are parallels, as witnessed by the transformation of socio-technical inspired Quality of Working Life programs into Japanese-informed Total Quality Management initiatives. Both strategies share an affinity for team structuring, payment-for-knowledge, and peer evaluations, as well as for job rotation and the utilization of cross-functional occupational groups.[72]

With fewer qualifications than the earlier proponents of responsible autonomy, contemporary advocates of post-Fordist work relations argue that the adoption of such practices have significantly beneficial effects in reversing the negative consequences that have been historically associated with scientific management. For Womack et al., lean production systems, as they refer to them, move us beyond Fordism, in so far as they use less of everything: 'half the human effort in the factory, half the manufacturing space, half the investment in tools, half the engineering hours ... far less than half the needed inventory on site.'[73] One outcome of this is the production of work that 'combines the advantages of craft and mass production' through the employment of teams of multiskilled workers. Accordingly, 'Most people – including so-called blue-collar workers – will find their jobs more challenging as lean production spreads.'[74] This is seemingly woven into the labour process of lean production systems, as a 'creative tension' which gives the system its productive traction. 'This creative tension involved in solving complex problems is precisely what has separated manual factory work from professional 'think' work in the age of mass production. ... by the end of the century ... lean assembly plants will be populated almost entirely by highly skilled problem solvers whose task will be to think continually of ways to make the system run more smoothly and productively.'[75] Owing to its superiority on both efficiency and sociological counts,

Womack et al. fully expect lean production to spread out to all industries in time.

A more scholarly account of the same phenomena is provided by Kenney and Florida.[76] Unlike Womack et al., their consideration of post-Fordist work relations extends beyond the auto industry to include steel, rubber, and electronics, as well as car parts and assembly. The key to the emergent labour process is what the authors term an innovation-mediated production system. This transforms the workplace into a laboratory where innovation and production, intellectual and manual labour are thoroughly integrated into an engine for continuous improvement. Thus, 'In the factory as laboratory, the distinction between intellectual and physical labor that Marx indicated and Harry Braverman raised to the fundamental contradiction of modern capitalism is at some fundamental level mitigated.'[77] Although acknowledging an intensification of work in enterprises that are ordered around the new labour processes, the authors insist that the 'cornerstone of innovation-mediated production lies in the harnessing of workers' intelligence and knowledge.'[78] This represents a fundamental reversal of the tenets of scientific management, including the debasement of labour thesis. But unlike post-industrial approaches, such outcomes are not premised upon the use of automated, continuous process technologies. Rather, they are the product of specific organizational changes that are dedicated to realizing an ethic of continuous improvement. In principle this is compatible with a range of technological menus.

What type of workers do the new social relations of post-Fordism require? Abstract, mathematical reasoning abilities are increasingly utilized on the shop floor, so that 'narrow skills must be accompanied by broad knowledge and an ability to understand and think abstractly ... to continually grasp new concepts.'[79] Indeed, for Kenney and Florida, post-Fordist workers as a whole increasingly resemble super-technicians, 'who monitor, review data and adjust and control the process ... These super-technicians have skill levels that are equivalent to electrical engineers of two decades ago' and must now 'be trained and managed more like a researcher than as a traditional factory worker.'[80] In addition to the enhanced analytical abilities that were not required under Taylorism or Fordism, new workforces will also have to be 'multiskilled.' By this, the authors are referring to the different layouts that often accompany post-Fordist reorganization. Instead of doing one job on one piece of equipment and then sending it on, workers may be required to tend a variety of machines in work station layouts that were first designed by Toyota engineer, Ono Taiichi.[81] 'Workers thus perform a number of tasks on different machines simultaneously while individual machines 'mind' themselves. Multiskilling is absolutely essential for this strategy to be successful.'[82] This is one version of the future that is now upon us, however, critics working out of a redeployed labour process perspective lend us a different picture, contrasting workers cum reskilled technicians with a different post-Fordist reality that is associated with lean production and Japanese production relations.

Burawoy, for example, argues that the hegemonic industrial relations of mature Fordism may be giving way to a new form of hegemonic despotism, in which capital mobility and the threat of capital flight are used to pry concessions back from labour in return for promised employment continuity and corporate financial success.[83] He does not pursue the concept of hegemonic despotism back into the labour process however. His comments on the topic are chiefly directed at the changing nature of industrial relations and factory regimes. Thus, his conclusions might be equally applicable to the radically marketized and spatialized relations that characterize late Fordism, and to the social relations of post-Fordism. For that reason, analyses which more adequately distinguish between the two are more useful.

In this vein, Dohse et al. have applied the term, 'Toyotism,' to the new labour processes that they see as being associated with post-Fordism. At one level, these labour processes are not so new. Rather '"Toyotism" is simply the practice of the organizational principles of Fordism under conditions in which management prerogatives are largely unlimited.'[84] The authors are referring to the intensification of work effort that Toyotism makes possible through the use of more broadly defined jobs, the reduction of staffing and inventory buffers, and the use of peer pressure. The introduction of these innovations was originally aided by the defeat of independent trade unionism in the Japanese automobile industry. The weakening of western labour movements through the mass unemployment of recent decades is in turn conducive to the transplantation of such production relations to other centers of accumulation. In this scenario, Ono Taiichi, Toyota's chief engineer, is credited with being Ford's logical successor in what, according to Price, is tantamount to a new form of 'intensified Fordism.'[85]

Similarly, Stephen Wood has argued that not all dimensions of so-called Japanese work relations represent a negation of either Taylorism or Fordism.[86] A more accurate portrayal according to Wood is to be found in the notion of hybrid models, which include important components of Taylorism and Fordism along with new innovations inspired by the success of major Japanese corporations. Although Wood does not get into a systematic assessment of what he terms 'neo-Fordism,' others writing from a critical perspective have. Berggren considers alternatives to the dehumanizing assembly line of Fordist industry, and clearly Japanese models of lean production do not qualify as satisfactory replacements in his evaluation.[87] As he argues, along similar lines to Dohse and Price, 'If anything, the rhythm and pace of the work on the assembly line is more inexorable under the Japanese management system than it ever was before.'[88]

Although Japanese production relations may well imply advances in productivity, this does not necessarily translate into superior working conditions for employees. For Berggren there is simply no fundamental break with Taylorism. In the Japanese automobile assembly plants that he inspected, work continues to be characterized by assembly line technology, short job cycles, and short training periods. Added to this are increased responsibility loads in the form of wider job areas (multiple machine tending), and the integration of responsibility for quality control into job descriptions. In an interesting twist, Berggren goes on to argue that for some functions, Toyota-style work relations actually represent an extension of Tayloristic principles. The successful emulation of small-batch flexibility in what remains a mass production industry required subjecting the tasks of retooling production lines to the maxims of Taylor.[89] 'The labor process in the Toyota system is thus designed according to classic Taylorist principles ... inventory less small-batch production requires a work force that is not highly specialized, but rather is capable of performing a number of different tasks. These different tasks ... are mainly variations of similar simple jobs.'[90]

This underscores a common theme. Post-Fordist production involves an intensification of work effort, but not necessarily an upgrading of labour forces. For Berggren, workers in post-Fordist environments alternate between similarly repetitive, standardized tasks, an activity that is better described as 'multitasking' rather than multiskilling.[91] Other authors have come to similar conclusions. In longitudinal survey analysis at one Japanese car assembly transplant, Robertson et al. report work intensification accompanying continuous improvement activity. For the authors, Tayloristic principles remain dominant and the outcomes associated with scientific management remain in effect. A Tayloristic standardization of work underwrites, as it were, the new forms of flexibility.[92] Parker and Slaughter have coined the term 'management by stress' to encapsulate the main features of these plants.[93] Among other characteristics, this refers to speeding up line flows, cutting labour and material buffers, and assigning employees greater levels of spatial responsibility. Parker and Slaughter consider it to be a misnomer to equate these trends with skill enhancement. As they argue, 'multiskilling every worker means deskilling every job.'[94]

Neither is the evidence compiled by Kenny and Florida in their cross-industrial examination of Japanese production relations compelling. Their study of a variety of Japanese transplants in the United States indicated that training periods still amounted to only one week per employee.[95] Job rotation seemed to be employed as much to relieve boredom and repetitive stress injuries, as to reskill employees, and make the need for job standardization and predictability that much greater.[96] At another Japanese transplant visited by the authors, workers spent three minutes at the start of each shift discussing continuous improvement/Kaizen, forty-five minutes per month on quality issues, and made biannual presentations to management on these issues.[97] These are not particularly impressive statistics.

As implied here, to date the greater part of the research that has been conducted on post-Fordist work relations has been preoccupied with a single sector, taking the automobile industry as the paradigmatic case. Although this has helped to focus the debate between proponents of post-Fordism as a desirable alternative, and sceptics of the new work relations, it has also unnecessarily restricted the debate by inadvertently controlling for industry. On the few occasions where there has been crossover into other sectors of the economy, similar issues have been uncovered. Thus, although the Japanese are credited by Kenny and Florida with creating a 'new iron age' in the American steel industry and with turning the mill into a laboratory for experiments in continuous improvement, a far different picture emerges from Corman et al.'s study of work reorganization at the Canadian steel giant, Stelco Ltd. Here work restructuring has entailed job amalgamation for production operators and multicrafting amongst the trades. This has translated into greater job discretion for workers, but along with that have come heavier work loads. The authors of the Stelco study are dubious about whether there has been much overall skill upgrading.[98] Recent changes in the industry are analysed over and against a dynamic of 're-Taylorization,' by which the authors are referring to periodic attempts to redesign the labour process in aid of ongoing, profitable valorization.[99] Although this is an excessively broad notion of Taylorization, there is no mistaking the broader trends in working conditions that the authors flag in their analysis. As in other criticisms of post-Fordism, work intensification, or doing more with less, is a common theme that runs throughout their study.

In a way this brings us back full circle. Post-industrialists pointed to a more attractive future, as undemanding, standardized jobs in industry were replaced by more challenging, holistic employment opportunities in the service sector and what was left of the industrial economy. The transformation was both occupational (that is, new jobs in new sectors) and socio-technical, as new labour processes embodying continuous flow production systems were adopted. These contentions were strongly disputed by a first generation of labour process theorists who argued on behalf of the staying power of Taylorism. Currently the terms of the debate have been recast; the focus is now much more on alternative paradigms for organization of the work-effort bargain and the claims that are made on their behalf. Still, old chestnuts remain in the fire. This is evident in the very terms of the debate – post-Fordism, re-Taylorization, work intensification, and multiskilling. By examining an industry that simultaneously employs different technologies in various phases of the production process and by analysing the impacts of different organizational choices made by major firms in that industry, we may be able to arrive at a clearer perspective on the sets of competing claims that have been outlined here.

Some Evidence on Skill and Related Matters

This section focuses the discussion on two related threads and attempts to progressively unravel them, prior to tying them back together again. One deals with corporate or managerial strategies and their effects on the labour process. This is the Fordism/post-Fordism debate. The other takes up the issue of occupational impacts on the labour process. This is the post-industrial debate. Of course, the two are indeed interwoven in history and practice. But momentarily distinguishing between them allows testing for both post-Fordist and post-industrial influences on the labour process and the question of skilling. Analysing any contrasting outcomes in the conditions encountered by workers at four quite distinct companies uncovers the effects that divergent managerial strategies may hold for the labour process and for reskilling and deskilling trends respectively. Analogously, by studying the experiences of the different occupations associated with mine/mill complexes, I am able to hone in on the claims of the post-industrial theorists. Because these labour processes are divided between various levels of mechanized mining activity and fully automated milling techniques, we can concentrate on the arguments that have been put forward by the post-industrial school through an examination of differences in the work experiences of miners, mill operators, and trades people.

Post-Fordist Influences?

What differences does working for one company in this study as opposed to another make, when it comes to considering the labour process and associated skill trends? According to one set of expectations, employment in a post-Fordist environment may be expected to utilize work practices that lead to the creation of more autonomous, skill-enhancing jobs, wherein workers exercise greater levels of discretion and control over the labour process. From a different perspective, post-Fordism is equated with the 'more with less' dynamic of contemporary political economy. Higher work expectations and the associated pressures, overshadow qualitative improvements in the conditions of employment. Here, the issue of whether or not any such expectations are borne out, over the range of managerial practices that have been elucidated in the preceding chapters, is addressed.

From what has been discussed, it would be reasonable to expect greater levels of workforce flexibility in post-Fordist plants. This could assume the form of greater levels of task diversity through practices such as systematic job rotation, the utilization of fluid job demarcations to permit the use of labour when and where it is needed (that is, flexible job assignments), and the expansion of job responsibilities. Tables 5.1 to 5.3 provide some important evidence on these matters by examining work practices at the five production sites. As indicated in the first table, a majority of those workers interviewed participate in job rotation schemes only at Agrium, where 51.4 per cent of the company sample indicated involvement in such schemes. Although Agrium workers constitute only 22 per cent of the sample, they accounted for 28 per cent of all those who regularly trade off jobs with peers, meaning that they are somewhat over-represented when it comes to the practice of job rotation (not shown).[100] This is reflected in the statistically significant, but not particularly robust, measure of association between employer and job rotation that is shown in Table 5.1 (Cramer's $V = .17$). Job rotation is especially common for both miners and mill operators at this mine. Given what we know about managerial policies at each site, this is hardly surprising.

Deliberate efforts at 'multiskilling' through rotational schemes are also in effect, but only selectively, at the Key Lake mill, where two-thirds of the mill subsample indicated participation in a job rotation scheme (not shown). Here, the exigencies of worker turnover and attendance problems associated with the long-distance transportation of the workforce to the site of production, along with the intention to keep the mill operating after mining operations close, have given management an immediate interest in enhancing labour force flexibility in this one area. As a result, job rotation at Key Lake is primarily a managerial initiative, but it is restricted to mill operations. In the mining department, workers are tightly controlled with respect to the types of equipment that they are allowed to operate. Without the requisite training hours, for example, workers are not permitted to operate cranes over a certain weight, or other pieces of equipment of given dimensions. Often, under the pressures of production, it is not possible to get the number of training hours in that would permit job rotation, even though workers have expressed an interest in it.

In the potash mines, on the other hand, much of the training that has been put into place has been of an *ex post facto* nature. Typically, mine and mill operators have been doing their jobs and switching places for years before formalized training became a part of corporate operations and record keeping. As a pre-existing aspect of working life in these mines, job switching is a practice that management is generally happy to see develop, with, or as in the case of PCS-Allan, without formal sponsorship. For in addition to potentially alleviating job boredom among workers, job rotation has provided management with an

TABLE 5.1
Job rotation

Job rotation		Allan	Central Canada	Cory	Cameco	Agrium	Row totals
			Mine site				
Yes	N	32	26	14	24	37	133
	%	47.8	35.1	27.5	35.8	51.4	40.2
No	N	35	48	37	43	35	198
	%	52.2	64.9	72.5	64.2	48.6	59.8
Column	N	67	74	51	67	72	331
total	%	20.2	22.4	15.4	20.2	21.8	100.0

Cramer's $V = .175$; $p = 0.39$.

inexpensive form of in-house, peer-based training, as well as with the potentially greater levels of workforce flexibility that flow from it. Each of these factors has come into play at the Allan mine, where management does not have a formal policy on job rotation. Nonetheless, this site is the other locale which is over-represented in the practice of job rotation, with 24 per cent of the reported rotation, but only 20 per cent of the sample. In this case, the practice is especially widespread among the miners, with 68 per cent of this workforce sub-sample engaging in job swapping activities, although management maintains no 'official' policy on the practice. It is an issue that is left up to each crew. Thus, miners at PCS-Allan engage in a good deal of informal job switching, but mill operators at the same mine do not. Here, job rotation would seem to have more to do with an autonomy that miners enjoy in their work, than with post-Fordist managerial commitments. Yet, along with the post-Fordist Agrium, Allan is over-represented in the category of job rotation, exhibiting similar overall proportions to Agrium in this practice (51.4 per cent versus 47.8 per cent).

Another element that is commonly identified with post-Fordist flexibility and workforce reskilling involves job reassignments to related parts of the labour process. Generally, these will be of short duration and, unlike job rotation, are initiated solely by management. Participation in these practices is usually seen to imply the exercise of multiple skills and hence the creation of a multiskilled, flexible workforce, where one worker can easily fit into the place and fulfill the functions of another. To capture this aspect of work experience, workers were asked whether they were ever temporarily reassigned to other jobs, job categories, or areas of the operation. Reassignments associated with lay-offs and bumping rights, a common feature of the seasonal employment regime at PCS-Allan, were eliminated from the comparison, as these were framed by a differ-

TABLE 5.2
Use of job reassignments[a]

Job reassignments		Mine site					Row totals
		Allan	Central Canada	Cory	Cameco	Agrium	
Yes	N	11	13	20	14	29	87
	%	25.6	25.5	50.0	23.3	46.0	33.9
No	N	32	38	20	46	34	170
	%	74.4	74.5	50.0	76.7	54.0	66.1
Column	N	43	51	40	60	63	257
total	%	16.7	19.8	15.6	23.3	24.5	100.0

Cramer's $V = .239$; $p = .005$.
[a]Does not include job reassignments resulting from inventory control and temporary layoffs.

ent agenda than flexible 'multiskilling' and would, by definition, usually involve moving into lower-skilled jobs.

The findings, contained in Table 5.2, reveal that when it comes to routine job reassignments at the behest of management, the results, in the first instance, are as mixed as they were in the case of job rotation. This practice was most common at PCS-Cory, where exactly half of the labour force sample indicated that they were frequently asked to undertake tasks other than their principal job by management. More predictably, Agrium was the next most common site for this practice. Both Cory and Agrium respondents are over-represented when it comes to experiencing job reassignments at the behest of management. Cory workers constitute 16 per cent of the sample, but 23 per cent of the positive responses to the question on job reassignments, while the Agrium workers made up 25 per cent of the sample and 33 per cent of the positive responses to this query. The relationship in this instance, between employer and job reassignments is considerably stronger than it is for job rotation and again statistically significant (Cramer's $V = .24$).

As noted above, frequent task reassignment not associated with lay-offs or bumping rights is often assumed to imply a process of multiskilling. This has been part of Agrium's managerial strategy, aided by the broad payment/job classification grid that is part of the company's history. At PCS-Cory, on the other hand, task reassignment has less to do with immediate managerial paradigms than with the effects of workforce downsizing. In short, fewer workers are called upon to do a wider variety of jobs. This goes hand in hand with the high levels of overtime that are also worked at this mine. Although these results

TABLE 5.3
Job expansion

Job expansion		Mine site					Row totals
		Allan	Central Canada	Cory	Cameco	Agrium	
Yes	N	39	54	30	39	53	215
	%	58.2	73.0	58.8	58.2	73.6	65.0
No	N	28	20	21	28	19	116
	%	41.8	27.0	41.2	41.8	26.4	35.0
Column totals	N	67	74	51	67	72	331
	%	20.2	22.4	15.4	20.2	21.8	100.0

Cramer's $V = .155$; $p = .092$.

should not be completely unexpected, they do have some interesting theoretical implications. Unlike Agrium, management at PCS has little time for new work organization paradigms. Nevertheless, the adoption of radically marketized industrial relations has produced very similar results in workforce flexibility, but without the jargon of continuous improvement and employee involvement, or financial supports such as gainsharing, which are developed in aid of such pursuits. This suggests that the importance of managerial adherence to new production philosophies might need rethinking. Clearly, there is greater flexibility in the deployment of workers at PCS-Cory and at Agrium. At the same time, these sites do not have much else in common. Conscious regulation by market forces, as is most evident in the radically marketized industrial relations at Cory, would seem to override many other considerations.

Job expansion, on the other hand, is a more common feature of employment at all of the companies (see Table 5.3). Identical amounts of task expansion were reported at Agrium and Central Canada Potash (73 per cent), which led the way in reported job expansion. At these mines, practically three-quarters of the workforce samples cited additions being made to their principal jobs since they had been incumbents in them. At each of the three remaining mines, over half of the sample (58 per cent in each case) viewed task addition to be a feature of their work. The existence of job expansion at workplaces that exhibit post-Fordist production relations should come as no surprise. Indeed, the ordering of the firms in Table 5.3 is in line with the expectations of both adherents to and critics of post-Fordist practices. However, the differences with more traditional firms are tempered by the fact that the variations are not significant at a statistically appropriate level, nor is the strength of the association between company and job expansion particularly noteworthy. What is perhaps more impressive is

TABLE 5.4
Reported levels of staffing

Staffing levels		Allan	Central Canada	Cory	Cameco	Agrium	Row totals
				Mine site			
Just about	N	33	24	8	36	37	138
right	%	49.3	32.4	16.0	55.4	51.4	42.1
Too many	N	1	2	0	1	5	9
workers	%	1.5	2.7	0	1.5	6.9	2.7
Not enough	N	33	48	42	28	30	181
workers	%	49.3	64.9	84.0	43.1	41.7	55.2
Column	N	67	74	50	65	72	328
total	%	20.4	22.6	15.2	19.8	22.0	100.0

Cramer's V = .23; p = .000.

not the differentiation between firms, but the high amounts of reported task expansion across all of the operations. It is an experience that is more common than not, irrespective of current managerial allegiances.

Similar results are revealed across a host of other indicators, with either an absence of significant variation between the mines, or some surprising clusterings that refuse to yield a clear distinction between the nominally Fordist and post-Fordist operations. For example, one hallmark of new production relations is teamwork. Yet, healthy majorities at each of the properties indicated working in a team environment.

By about the same margin, large majorities at each mine reported exercising complete control over the pace at which they worked. Probed further, workers at PCS-Cory admitted to being rushed in greater relative numbers than anywhere else (22 per cent of the sample) followed by PCS-Allan at 10 per cent. At the other mines 5 to 7 per cent of those interviewed indicated that they worked under pressure. Cory workers, who have been subjected to the most intensive downsizing are, in particular, over-represented in this category. Constituting 16 per cent of the sample, they represent over a third (36 per cent) of the respondents who reported feeling rushed for much of the time in their work. On the other hand, workers at Agrium and Central Canada, the two operations most closely approximating post-Fordist ideals, were the least likely to report working under rushed conditions and were under-represented among the respondents.

The same point is registered with respect to levels of supervision, understaffing, and management production norms. Flexible, post-Fordist, lean production has been identified with downsizing in the ranks of both hourly production

TABLE 5.5
Reported production norms

| Work expectations | | Mine site | | | | | Row totals |
		Allan	Central Canada	Cory	Cameco	Agrium	
Too high	N	6	12	11	3	4	36
	%	12.2	24.0	39.3	7.1	8.9	16.8
About right	N	42	33	17	37	36	165
	%	85.7	66.0	60.7	88.1	80.0	77.1
Too low	N	1	5	0	2	5	13
	%	2.0	10.0	0	4.8	11.1	6.1
Column total	N	49	50	28	42	45	214
	%	22.9	23.4	13.1	19.6	21.0	100.0

Cramer's $V = .24$; $p = .002$.

workers and supervisory staff. Be that as it may, the two firms that are clustered together on both of these variables are Central Canada Potash and PCS-Cory. Near identical majority proportions at each site (55 and 57 per cent, respectively) indicated declines in the level of supervision over their work, while at the other mines, majorities reported constant levels of supervision.

Similarly, at Cory and Central Canada, majorities of the sample (84 and 65 per cent, respectively) considered that there were not enough workers on site, given the amount of work to do (Table 5.4). At the other three mines, majorities considered the workforce numbers to be 'about right.' In both instances, the results pertaining to levels of supervision and production staffing were highly significant with strong levels of association between the mine site and the responses to the questions on these issues.

With respect to work expectations, again workers at Cory and Central Canada, in that order, recorded in significantly higher proportions that such production norms were too high (Table 5.5). At Cory, almost 40 per cent of the sample gave this response, constituting nearly one-third (31 per cent) of all workers who looked upon production targets as too high. Similarly, at Central Canada, one-quarter of the sample concurred with this judgment, representing exactly one-third of the total number of workers who considered production norms to be excessive. Both Cory and Central Canada were clearly over-represented by workers who thought that labouring expectations were too high, while the other sites were under-represented by respondents in this category.

Returning to the original query regarding the impact of different managerial strategies for the elicitation of work effort on skill trends, these results indicate

TABLE 5.6
Skill trends

Skill requirements		Mine site					Row totals
		Allan	Central Canada	Cory	Cameco	Agrium	
Decreased	N	8	3	4	2	1	18
	%	13.1	4.3	8.0	3.2	1.4	5.8
Stayed the same	N	14	13	17	24	21	89
	%	23.0	18.6	34.0	38.1	30.4	28.4
Increased	N	39	54	29	37	47	206
	%	63.9	77.1	58.0	58.7	68.1	65.8
Column totals	N	61	70	50	63	69	313
	%	19.5	22.4	16.0	20.1	22.0	100.0

Cramer's V = .17; p = .02.

some variation between the companies included in the study. Agrium and Central Canada's prominence in a number of these categories (job rotation, work reassignments, job expansion, decreasing supervision, and understaffing) might be expected theoretically, but the results are immediately tempered by the high-placed inclusion of more traditional firms such as the Cory and Allan divisions of PCS in some of the same signifiers. What stands out, then, is not so much the variation between our nominally Fordist and post-Fordist producers, but a commonality of experience that is grouped around employment shrinkage and the 'more with less' phenomenon. This holds true regardless of whether or not the concerns in question are ostensibly Fordist or post-Fordist in nature and is clearly seen in the clustering of both Central Canada and PCS-Cory around a number of response categories that have a direct bearing upon the labour process. With similar frequencies, workers at both companies reported common experiences with respect to declining levels of supervision, inadequate staffing levels, and high work expectations, in spite of ostensibly different managerial agendas.

This finding also seems to extend to the results on skill trends, proper, at the different mines. The picture that comes through in Table 5.6 is one of skill upgrading across the employer spectrum. Significant majorities at each company indicated increased skill demands in their jobs. Although the pattern of upgrading is somewhat stronger at the two firms that have ventured into post-Fordist alternatives, this feature is tempered by the reported upward revision in skill levels that is displayed across the study. Thus, Central Canada Potash and Agrium recorded the highest proportions of reported skill revision respectively

(77 and 68 per cent of these samples), but levels of upgrading were not grossly less at the other sites. No one firm is either seriously over- or under-represented in the skill enhancement category. Skill consistency is a secondary theme that emerges at Key Lake, Cory, and Agrium, while reported deskilling is of notable proportions only at PCS-Allan and PCS-Cory. Even here, however, it is a decidedly minority response. Overall, skill upgrading does seem to be related to the managerial practices that are exercised at the five production sites, but not in as bold a fashion as the proponents of post-Fordist alternatives might have expected. These results present both a puzzle and some clues for unraveling the mysteries of reported skill trends. Work practices that were thought to break down neatly along the Fordist/post-Fordist dichotomy and to therefore have a predictable impact on skill trends, do so, at best, in only a weak and unconvincing manner. On the other hand, employment reduction and the consequent fall-out were registered by all companies and seemingly lie behind such common practices as work assignment flexibility and job expansion. Into this cauldron enter the post-industrial arguments, which are considered here for what they can offer in accounting for the emerging trends.

Post-industrial Occupations?

It is conceivable that skill trends in the mining industry, as elsewhere, may be influenced by the incorporation of new technologies and occupations into the production process. With its divisions into mining, milling, and maintenance departments, modern-day mining activity presents a usefully diverse environment in which to critically examine the claims of post-industrial theory. As previously seen, mining occurs under highly mechanized conditions, while both potash and uranium milling are continuous process industries, in every sense of the word. According to some post-industrial theorists, workers in highly automated production processes would have more in common with craftworkers than with other production workers. As applied to this study, then, mill workers ought to be situated more closely to craftworkers (that is, maintenance personnel) than to miners in the nature of their work and in their responses to queries about it. Additionally, as mining activity itself takes on more and more of the characteristics of a continuous process industry, skill levels ought to be given a further boost. Here, my interest lies in operationalizing and pushing these claims to the empirical threshold.

One generally agreed upon indicator of skill is background training. In terms of actual training required to do the job, these workforces break down along traditional operator and trades lines. Operators, regardless of whether they work underground or in the mill, report training periods in terms of weeks, while trades occupants register training programs that take an average of four years to complete. This is important because amounts of training are often taken to be an indicator of skill gradient. Yet in response to questions pertaining to training, no significant variation in the amounts of formal training received by mine and mill operators amongst the complete sample were evident. In both instances, majorities reported training that lasted no longer than one week. With one exception, this pattern also holds for individual sites. At each, the largest proportion (although not necessarily majorities) of both mine and mill workers reported receiving no more than one week's formal training in order to do their current job. At the open-pit Key Lake mine, the largest proportion of mill workers (50 per cent) indicated training periods lasting up to two weeks, while 40 per cent reported periods of from two to six months and up to two years.

Familiarization, or the length of time before one feels comfortable that job expectations have been mastered, is different from actual in-house training, but this provides another indicator of required skill. Thus, the longer the familiarization period for any job, the more skill that is required, given that other things such as employee competence remain constant. On this count, there are significant differences between the experiences of mine operators and mill workers, with the latter reporting longer on-the-job learning curves. The largest proportion of miners (30.5 per cent) described familiarization periods of less than one week for the positions that they currently hold. The largest share of mill operatives (21 per cent) indicated that familiarization periods lasted from one to three months for their jobs. On the whole, mill workers were more likely to respond to questions about familiarization in terms of months of experience rather than weeks.

The same pattern holds across the individual companies and is outstanding in the case of Agrium, where fully 44 per cent of the mill operators reported familiarization periods of two years or more. Not coincidentally, Agrium is the one company that has a formalized job rotation system in the mill, but as well, miners at Agrium reported somewhat longer familiarization periods than elsewhere. This may reflect the generalized, broad banding of jobs at the company and the four-year-long operator trainee positions. Overall, mill operators have wider work areas than are encountered in underground mining, where labour tends to be associated with competence in using specific pieces of equipment. This may account for the longer periods of familiarization that are conveyed by the mill operators across the set of firms.

As in the previous section on the question of skill proper, respondents were asked whether or not they considered that the skill requirements for their jobs had decreased, remained constant, or increased since they had been doing them. The overall results are found in Table 5.7. Mill operators were second only to mine maintenance personnel in claiming skill upgrading, which may offer support for the post-industrial thesis describing the skill-enhancing effects associ-

TABLE 5.7
Skill trends by occupation at five worksites

Skill requirements		Occupation				
		Mine operator	Mine maintenance	Mill operator	Mill maintenance	Row totals
Decreased	N	8	1	4	5	18
	%	9.3	1.7	4.3	6.9	5.8
Stayed the same	N	38	9	21	21	89
	%	44.2	15.0	22.3	29.2	28.5
Increased	N	40	50	69	46	205
	%	46.5	83.3	73.4	63.9	65.7
Column totals	N	86	60	94	72	312
	%	27.6	19.2	30.1	23.1	100.0

Cramer's $V = .20$; $p = .000$.

ated with automated production processes. A considerably higher proportion of mill operators claim to have experienced upward skill revision than is the case with mine operators. But intriguingly, even amongst the latter group, the largest proportion, 47 per cent, reported the skill enhancement of their work, compared with 43 per cent who stated that skill requirements had remained constant and 9 per cent who reported deskilling effects. Overall, 73 per cent of the mill operators had seen the skill levels for their positions increase over time, while most of the remainder reported constant skill requirements for their work. As indicated, the association between occupation and reported skill trend is robust and non-casual.

These findings definitely complicate neat theoretical generalizations. That is, to the extent that mill workers report skill upgrading in greater proportions than miners and are slightly over-represented in the group of skill-enhanced workers, the expectations of post-industrial theory are borne out. But the extent of upgrading in the mining division itself gives pause for thought. In this still pre-automated work environment, what would explain such a reported skill trend?

In passing, these results also might seem to bear some affinity with the previous conclusions regarding Fordist and post-Fordist comparisons, where skill levels were seen to rise across the board, even if somewhat more frequently in the latter operations. In the occupational analysis, employees in the industry reported on skill accretion, although some, such as the mill workers, did so in higher proportions than others. On the surface at least, the findings are also completely at odds with the original predictions of critical labour process theory, which has focused on deskilling activities. On the other hand, these patterns bear

TABLE 5.8
Control of technology by occupation at five worksites[a]

		Occupation				
Level of control		Mine operator	Mine maintenance	Mill operator	Mill maintenance	Row totals
Fully control	N	69	40	46	53	208
technology	%	72.6	66.7	46.9	68.8	63.0
Partially control	N	19	16	37	21	93
technology	%	20.0	26.7	37.8	27.3	28.2
Controlled by	N	6	4	15	3	28
technology	%	6.3	6.7	15.3	3.9	8.5
Column	N	95	60	98	77	330
totals	%	28.8	18.2	29.7	23.3	100.0

Cramer's $V = .15$; $p = .009$.
[a]calculations include 1 missing observation.

a striking resemblance to the findings uncovered by Gaullie, discussed earlier, where the degree of labour power upgrading was directly related to the extent of automation in the workplace.[101] As mechanization and automation take hold, any workers who are situated in such environments are likely to report higher skill acquisition. In the cases examined here, however, this effect has little to do with company training programs. This vitiates the connection that Gaullie makes between automation, training, and skill level and still leaves the problem of explaining the reported skill trends that are exhibited in Table 5.7.[102]

Post-industrial theorists link these outcomes to the demands of new, computerized technologies. The automated work environment is reputed to require advanced, analytical problem-solving abilities for the purpose of 'controlling the controls.' But, when queried about their perceived relations with the technologies that they use in their daily work, mill operators reported the lowest levels of control. Majorities of miners and subsurface/surface maintenance workers claimed to exercise full control over the technologies that they employ. Of the mill operators, however, 47 per cent claimed to be in full control over the technology, 38 per cent in partial control and 15 per cent not in control over the immediate forces of production (Table 5.8). Although this finding may not be completely incompatible with post-industrial arguments, it does cast the concept of skill in a new light. Skill has usually been associated with mastery of a task or process through the adept use of specific tools or forces of production. In this case, however, it may be the relative sense of an *absence* of mastery or control, or the very contingency of such control and the associated *sense of risk* around the use of capital intensive technologies, which accounts for the reported skill-upgrading effects. Thus, although the ideal of post-industrial theory may be mastery of complex new technologies by highly skilled workers 'controlling the controls,' at present the reality may well be something less than that. This interpretation is also suggested by the somewhat longer familiarization periods reported by mill workers. Whether this is a transitory state, or part of a new post-industrial reality, is another matter.

Probing further into the labour process reveals other conflicting trends. For example, questions on control over the pace of work yielded almost identical results from mine and mill workers. Although majorities in both groups indicated that they usually or always control the pace of their work, the proportions were significantly lower than for the tradesworkers. At all the sites, then, mill workers had more in common with miners than with tradespeople in the level of their control of pace of work.

The same pattern was found in answers to queries on opportunities to use personal judgment in the way a job is carried out. This is one indicator of job autonomy, and on it, almost identical proportions of mine and mill workers indicated considerable, some, and little room for the use of judgment in their jobs. For example, 12 per cent of the miners and 8 per cent of the mill operators reported having little room to use their judgment on how to do their work. This compared with 3 per cent of the mine and 1 per cent of the mill maintenance workers who stated that they had little opportunity to use their own discretion in carrying out their work. Again then, both operator groups fell behind the trades in reported potential for use of personal judgment on the job.

Exploring autonomy in the labour process further, mill workers seem to have more discretion than miners in deciding the order in which they will do the tasks that they are responsible for. In their ability to make these choices, the mill operatives as a group fall between the tradesworkers and the miners, with 70 per cent indicating that they have the autonomy to decide the order in which they do their work tasks, as compared with 53 per cent of the miners, 85 per cent of the mine maintenance personnel, and 79 per cent of the mill maintenance employees.

Task order is only one element of discretionary judgment however. Overall, underground and surface operators exhibit little significant variation in the potential to exercise judgment on the job. This was evident again, when employees were questioned about work norms. For the post-industrialists, workers in automated production settings are partially removed from the act of production, which proceeds at arm's length from the workforce. Work takes on an analytical quality as opposed to the manual connotations that it formerly exhibited. As such, the methods of Taylorism no longer have a place in the post-industrial work setting. Any form of drive system would be inappropriate, if not counterproductive.[103] Yet in these interviews, both mine and mill workers indicated an awareness of specific company work expectations, as translated into

TABLE 5.9
Participation in job rotation by occupation at five worksites

		Occupation				
Job rotation		Mine operator	Mine maintenance	Mill operator	Mill maintenance	Row totals
Yes	N	47	14	46	25	132
	%	49.5	23.3	46.9	32.5	40.0
No	N	48	46	52	52	198
	%	50.5	76.7	53.1	67.5	60.0
Column	N	95	60	98	77	330
total	%	28.8	18.2	29.7	23.3	100.0

Cramer's $V = .21$; $p = .002$.

optimal production targets that crews were expected to meet. For miners, so many tons of muck, or feet of wall cut per hour or shift, were the norm. For the mill operators, running a specific tonnage of muck through the mill was the operant expectation. Specific work expectations were more common among all classes of operators than among the trades. For the latter occupations, each job had an expected time allowance, but these were no more than loose guidelines. Major equipment overhauls or 'rehabs' are much like house renovations; the worker is uncertain of what a job will entail until well after work has begun, and management is aware of this contingency. Even routine maintenance overhauls can provide unexpected complications. In operations, on the other hand, although production quotas were not in effect in any of the mines, it was suggested that continued deviation from management production norms would meet with serious questioning. Thus, between mine and mill operators the differences regarding perceptions of managerial work targets were insignificant. Large majorities in each occupation (73 per cent of mill operators and 70 per cent of miners) were aware of corporate work expectations, as realized in production norms. For maintenance workers, 50 per cent of those working in the mine maintenance departments and 61 per cent of those in the mill maintenance divisions were consciously aware of specific managerial work targets. Consciousness of such definable expectations again serves to differentiate both mine and mill workers from their co-workers in the trades to a significant degree. Work norms play a more important role in the former cases.

In accordance with post-industrial scenarios and as an *explanans* for the skill trends reported upon in Table 5.7, we might expect the mill to be *the* site of new working relationships. This is not necessarily the case, however. For instance, contrary to theoretical expectations, the mills are not the most common areas

TABLE 5.10
Job reassignments by occupation at five worksites[a]

Participation in job reassignments		Occupation				
		Mine operator	Mine maintenance	Mill operator	Mill maintenance	Row totals
Yes	N	31	8	32	16	87
	%	44.3	17.8	42.7	24.2	34.0
No	N	39	37	43	50	169
	%	55.7	82.2	57.3	75.8	66.0
Column total	N	70	45	75	66	256
	%	27.3	17.6	29.3	25.8	100.0

Cramer's V = .23; p = .003.
[a]Does not include job reassignments resulting from inventory control and temporary lay-offs.

for job rotation. In fact, Table 5.9 reveals that such practices were marginally more common amongst miners, half of whom regularly trade off jobs, compared with a slightly smaller proportion of mill workers, (47 per cent). Again, however, the main point of variation is between operations and trades. Job rotation is much less frequent among the trade groups, both underground and on surface (23 and 33 per cent of mine maintenance and mill maintenance workers respectively). Between miners and mill operators, on the other hand, and across the sample as a whole, there are no significant differences in the levels of job trading, except at Key Lake, which has formalized job rotation in the mill, while prohibiting it in the pit. What we witness, then, is not a convergence between craft and automated production processes, but continued differentiation. Mine and mill workers are clustered together around the practice of job rotation, but are clearly differentiated from the crafts on this count.

Again, higher levels of multiskilling might be expected in the mills, owing to the holistic, arm's-length nature of the labour process referred to by post-industrial authors. But this expectation is not realized either. As with job rotation, significant variation in the experience of being subjected to job reassignments exists between operations and trades, but not within the operator classes themselves (Table 5.10). Thus, across the sample as a whole, 44 per cent of mine operators and 43 per cent of mill workers reported that task reassignment, unrelated to lay-offs, is a part of working life. This is about double the incidence given by the tradesworkers who were queried, and accounts for the bulk of job reassignments. The only significant deviation from this result is recorded at Agrium, where 87 per cent of the mill sample is frequently given job reassignments as opposed to 55 per cent of the miners. Given the overall pattern, however, this variation has more to do with the company and its policies than with occupational differentiation per se. As in job rotation, the use of flexible work assignments involves greater proportions of miners *and* mill operatives than trades personnel. This result makes it that much more difficult to use such practices in explaining differing skill trends between the groups. It should also give us pause for thought when we treat skill as a singular, homogeneous quality that can be possessed in greater or smaller amounts.

Teamwork is also often taken to be an attribute of post-industrial realities. Among all of the occupational groups, miners were the most likely to indicate involvement in work teams. To a large degree, this is a function of the physical nature of mining, which is undertaken as a joint activity by a small group of workers. That just over 88 per cent of the total sample of miners indicated involvement in group work is thus no surprise. What is more interesting, however, is the finding that mill operators ranked last in work group participation, with 63 per cent of the total sample of mill operators stating that they regularly worked as part of a team. The organization of this aspect of the labour process, with its extreme capital intensity, seemingly favours more-isolated work experiences where one control-room operator remains in periodic radio contact with one floor operator over the course of a shift.

On a number of important dimensions that are associated with the organization of work and that might logically be connected with reported skill trends, the main variation continues to be between operators as a job class and tradesworkers. This same division also appears to be the most salient one in the ongoing micropolitics of each local that took part in the study. Meanwhile, differences have not emerged between miners and continuous process workers around multiskilling activities, job rotation, control of pace of work, participation in teamwork, room for the exercise of judgment, or the significance of company production norms to the extent predicted by post-industrial theory. Process operators, as a whole, do express a less certain attitude with respect to the technologies that they use than do miners, which may imply more demanding learning curves (longer familiarization periods) and higher reported skill accretion. As well, mill operators do indicate higher levels of discretion over the order in which they do their job than is the case for miners, although both groups specify almost identical capacities for the use of self-judgment in the labour process. In short, the similarities between miners and mill workers along a number of consequential dimensions outweigh the differences. Where there are differences in a post-industrial direction, they sometimes favour the mill operatives, but just as often, highlight the position of the miners.

There is one other aspect of work organization, related to those discussed above, that does go some way in accounting for the skill trends that were reported on earlier. When queried about job expansion, all occupations recorded its

TABLE 5.11
Job expansion by occupation at five worksites

		Occupation				
Job expansion		Mine operator	Mine maintenance	Mill operator	Mill maintenance	Row totals
Yes	N	53	38	73	50	214
	%	55.8	63.3	74.5	64.9	64.8
No	N	42	22	25	27	116
	%	44.2	36.7	25.5	35.1	35.2
Column	N	95	60	98	77	330
total	%	28.8	18.2	29.7	23.3	100.0

Cramer's $V = .15$; $p = .058$.

occurrence, but as Table 5.11 shows, this feature is a more pronounced in continuous process milling operations than elsewhere.

Across the sample included in this study, mill operators more frequently chronicled job expansion, that is, tasks being added to their principal job, than did their counterparts elsewhere in the occupational structure of the industry. In fact, on this point, there is greater variation, and at a higher level of significance, among the operator classes (mine and mill workers) than between them and the trades. Although the relationship that is shown in Table 5.11 between the experience of job expansion and occupation is moderate and just significant at the .05 level, when the comparison is restricted to mine and mill operatives alone, the relationship between occupation and job expansion is stronger (Cramer's $V = .2$) and highly significant ($p = .006$). At every site, mill employees proportionately outnumbered miners in citing job expansion as a part of working life. At two of the mines, Agrium and Key Lake, mill operators topped the occupational list in reporting job expansion, while at Central Canada Potash and PCS-Cory, there was a 1 per cent difference between the percentages of mill workers and mill mechanics who were effected by job expansion. Overall, three-quarters of the mill operatives reported experiencing task expansion, compared with 56 per cent of all the miners. Maintenance workers fell between mine and mill operators, while mill operators stood alone as being clearly over-represented in this facet of employment. And, in nearly all instances, the initiatives to reorder work by expanding tasks were undertaken unilaterally, at the behest of management. Rarely, if ever, were those most immediately affected by such alterations consulted beforehand or allowed input into the expansion of their responsibilities.

Related to task expansion is the issue of staffing levels. When asked about the adequacy of current employee rosters given the amount of work to do, a majority of mill workers (54 per cent) observed that there were labour scarcities in their work areas, compared with 48 per cent of the miners. Underground maintenance personnel reported understaffing in approximately identical proportions to the mill workers, while 70 per cent of mill maintenance workers agreed that labour scarcities existed in their areas. This shortage may manifest itself in different ways depending upon the job. Maintenance workers, for example, sometimes mentioned the lack of what they consider to be adequate upkeep at the workplace. Instead of making repairs for the longer term, faster, short-term solutions may prevail. Thus, while workloads do not necessarily increase, some workers express dissatisfaction at what they are *not* allowed to do, or at what they consider to be a lack of proper maintenance. Operators, on the other hand, are prone to connect falling employment numbers with larger work areas. Again, however, this does not necessarily imply heavier workloads. The latter may be offset by improvements in technology that reduce the amounts of 'bull' work.

Conclusions

Thus far the findings have produced the following curious paradox. Post-Fordist and post-industrial claims around the future of work have been upheld, but in a rather weak sense, in this analysis. Higher proportions of those working at ostensibly post-Fordist firms, or in post-industrial occupations, reported on skill-enhancing effects than is the case with their peers in Fordist or mechanized labour processes. *But,* and it is a critical qualification, these results do not seem to be occurring for the reasons given in post-Fordist or post-industrial arguments. The post-Fordist firms and the post-industrial occupations in our study show no more consistent propensity to adopt those practices that have been singled out as being responsible for the reskilling of labour than their counterparts. What then is behind the common trend pertaining to skill that runs across both occupational categories and employers' managerial strategies?

In order to respond to this query and to move beyond the bivariate relationships that have been considered thus far, several logistic regression models were created to try and advance the analysis. The most promising results are contained in Table 5.12. In it, the factors considered up to now have been entered as independent variables with the intention of accounting for reported skill trends. As this technique requires that the dependent variable assume a dichotomous status, skill level has been categorized as decreased/constant *or* increased. Given that a majority of cases report skill increase, and that this constitutes the paradox, the advantage of this operationalization is self-evident; it focuses attention on the most frequent outcome and compares it with the others.

The concern then lies with estimating the impact that several factors associ-

TABLE 5.12
Logistic regression estimates predicting skill trends at five worksites

Variable	B	Significance	Exp B
Occupation		.001	
[Mine operations]			
Mine maintenance	1.650	.000	5.20
Mill operations	.897	.015	2.45
Mill maintenance	.276	.484	1.31
Familiarization period		.246	
[<1 Week]			
1–2 Weeks	.158	.752	1.17
2–4 Weeks	.482	.348	1.62
1–3 Months	.316	.470	1.37
3–6 Months	.860	.089	2.36
6–12 Months	.072	.880	1.07
1–2 Years	1.77	.012	5.87
> 2 Years	.714	.246	2.04
Staffing levels			
[Just about right]			
Inadequate	−.212	.138	.809
Levels of supervision			
[Increased]		.629	
Stayed the same	−.547	.337	.578
Decreased	−.505	.384	.603
Relationship to technology			
[Controlled by technology]		.581	
Fully control technology	−.028	.954	.972
Partially control technology	.329	.542	1.39
Participation in job rotation	−.084	.771	.919
Job reassignments			
[due to lay-off]		.702	
No	.036	.921	1.03
Yes	.301	.474	1.35
Experienced job expansion	1.05	.000	2.89
Job lay-offs	.126	.669	1.13
Constant $N = 300$	−1.77	.119	

ated with post-Fordist work reorganization and post-industrialism, and already considered singularly, have on reported skill trends. The variables that are entered include occupation, length of time on the job before full familiarization has occurred, the perceived adequacy of current staffing levels, trends in the levels of supervision, perceived levels of control exercised over the technology in use, participation in job rotation, managerial use of job reassignments, job expansion, and finally, lay-off experiences. These entries will by now be famil-iar except for the last. Each factor is of considerable importance in either post-Fordist or post-industrial accounts of contemporary work experience or both. The final element has been added because lay-off experience does represent another distinction between the firms in this study. Workers who are frequently laid off may view this as disconfirming of their skill values, or alternatively, they may have less opportunity to see their skills validated.

In the 'Variable' column of Table 5.12, the category in square brackets is the referent value. The coefficients B and Exp.(B) are measures of deviation from the referent category. If Exp.(B) is greater than 1, then the odds are increased that the dependent variable, skill enhancement, will occur. If they are less than 1, the odds are decreased of skill revision taking place through the effect of the independent variable.

From Table 5.12, three factors emerge as having an independent impact upon skill assessments. The most significant is job expansion. Workers who are now expected to do more with less are more likely to report skill-upgrading effects. This does not mean that each individual task that has been added to a position is more demanding, but simply that in the aggregate, the job as a whole is such. As we have seen, job expansion has occurred across all the sites of production. Regardless of whether companies subscribe to post-Fordist work relations or more traditional formats, job expansion seems to be a commonality that has gone hand in hand with a reduction in employment numbers. For this reason, the employing company is not a particularly relevant factor and has no signifi-cant effect on reported skill trends. However, a common practice, multitasking, at all of the companies, does.

The other variables with significant independent impacts upon reported skill enhancement are the learning curves (familiarization periods), for the jobs and occupation. Positions that require over a year for full familiarization are likely to be evaluated as skill increasers apart from any job expansion that may be asso-ciated with them. In this regard, mill operators were more likely to report longer familiarization periods before the 'comfort zone' for a job is reached. But apart from this, occupational category also exerts an independent effect on skill. The referent category in this regard is mine operator. Although it should come as no surprise that trades membership would enhance skills, as compared with under-ground operator status, the same results also hold true for the mill operators. Com-pared with miners, work in the mill is more conducive to reporting skill upgrading. This result is independent of either the job expansion, the familiar-ization period, or any of the other variables that have been entered into the model.

Other factors are, statistically speaking, insignificant in perceptions and reports relating to skill. Thus staffing levels that are viewed as being inadequate do not have an independent bearing on the issue. Neither do decreasing levels of supervision, when they are contrasted with increasing levels. Finally, job reassignments that are not associated with lay-offs do not exert a significant effect on the skill question, nor for that matter does participation in job rotation.

Overall then, many of the hallmarks of post-Fordist work relations, as well as the companies that support them, fail to register any noticeable effects on the skill trends that are experienced by their subjects. Contrary to proponents of such relations, they are not convincingly connected to more challenging work in the manner in which they were intended. Rather as a more critical eye would have it, multitasking is of greater consequence in accounting for contemporary movements in skill requirements. It is the singular most important factor in accounting for the responses that have been collected on the issue of skills. Furthermore, this is principally related to the 'more with less' phenomenon – fewer workers performing a larger share of the collective effort – and it has occurred across the companies considered in this text, irrespective of the managerial discourses that they present to their workers.

Although others have drawn attention to the phenomena of 'multitasking,' the reality of it can too easily be dismissed. That it entails doing more with less also seems to indicate the imposition of new demands that workers equate, not incorrectly, with the exercise of greater skill requirements – a point that is frequently missed in the post-Fordist critiques. In sum, the work situation may indeed by more 'challenging,' but not for the reasons adduced by the proponents of post-Fordism. Accomplishing more with less places new demands upon workers, but this does not add up to the creation of new cadres of analytical problem solvers.

Although the results do not provide much support for the claims of post-Fordist advocates, more support is offered to the post-industrialists. Working in continuous process mills is equated with higher skill demands by the workers. The effects of this, independent of either task or learning curve expansion, need to be explored further. Although we have demonstrated a statistical relationship between continuous process work and reported skills, our theoretical understanding of this connection is still far from complete. Still, both post-Fordist and post-industrial effects in the workplaces examined in this study are overridden by the dynamic of the more with less phenomena. Highlighting this point is indeed the strong suit of contemporary labour process theory. This dynamic, as opposed to general claims about deskilling, which are much more difficult to verify, ought to be emphasized in a contemporary labour process perspective.

The next chapter shifts the focus to the industrial relations effects of alternative forms of work organization. There the influences of post-Fordist protocols on the political apparatuses of production and production politics at the five workplaces are scrutinized in greater detail.

6

Production Politics at Five Mine Sites

Along with the imputed skill effects that have been attributed to new forms of work organization, such practices are also claimed to support a better overall industrial relations climate at the workplace. Although skill trends and industrial relations are obviously interrelated, broader labour/management relations require assessment in their own right. In other words, industrial relations have a certain autonomy from labour processes, which must be taken into consideration in any analysis of workplace reorganization.

Canadian political economist H.C. Pentland provides a useful starting point in his characterization of industrial relations as those activities associated with 'marshaling the working population to get the necessary work done.'[1] Under this rubric, Pentland includes not only the supply-side questions of acquiring and retaining a labour force, issues that were dealt with in Chapter 2, but the whole question of coordinating, motivating, and disciplining workers to perform. For Pentland, 'The [industrial relations] system must do several things. ... It must co-ordinate and discipline the efforts of its labour force by a reasonably consistent and acceptable set of laws and customs, based on mores that command substantial consent. It must provide systems of rewards and punishments that produce effective motivation.'[2]

Several items are mentioned here, some of which, such as the coordination of effort, clearly overlap with the organization of the labour process. Others, however, entail appeals to laws and accepted mores that provide the basis for consent, as well as the ground rules for adjudicating conflict. This is the sphere of industrial relations proper, which other authors, along with Pentland, have identified with the 'rules governing employment and the ways in which they are made and interpreted.'[3] I will refer to this framework as the political apparatuses of production. The ways in which these rules are applied and creatively used by the participants to the labour process will be termed production politics.

The notion of political apparatuses of production, although likely less familiar to readers than the term 'industrial relations,' is a more precise concept. It refers not only to the rules that are used to regulate the effort bargain in the labour process but also to their inherently political dimension. Burawoy sums up these points nicely when he explains: 'the process of production contains political and ideological elements as well as a purely economic moment. That is, the process of production is not confined to the labour process – to social relations into which men and women enter as they transform raw materials into useful products with instruments of production. It also includes political apparatuses which reproduce those relations of the labour process through the regulation of struggles.'[4]

The political apparatuses of production may be viewed as political in two senses. First, as codified rules that are enunciated, for example, in collective agreements, grievance procedures, and disciplinary protocols, they represent non-economic resources that are subject to strategic deployment on the part of the participants to the labour process in the production politics they create. Often they have the backing of law, or quasi-legal status, as in the legally binding nature of a collective agreement, or in the precedents that are set by arbitrational rulings. Second, such protocols for governing the labour process have often been framed in broad strokes either by the state, at the behest of the state, or through state processes of interest mediation.

The latter dimension deals more with the origin of the political projects that have been designed for regulating the employment relationship. As this history has received extensive coverage elsewhere, here, the actual apparatuses of regulation will simply be taken as a background datum, which nonetheless is subject to all important strategic utilizations on the part of the participants to the labour process.[5] This chapter is more concerned with analysing the practices or production politics around the political apparatuses of production in specific worksites with different production paradigms. For example, workers at each mine site are covered by collective agreements governing conditions of work such as wage determination, job transfers, lay-off procedures, overtime, and the maintenance of safe working environments. Employer disciplinary codes and the grievance process are also part of the political apparatuses of production. Practices around the collective agreement, including the enforcement of discipline (means and rigour), the frequency with which recourse is made to grievance machinery, and the outcomes of such conflict constitute the essence of production politics proper. In other words, production politics entails an analysis of the *strategic use of the political apparatuses of production by the parties to the employment relationship*. This is the central concern here.

In the main the political apparatuses of production are broadly similar at each site. Each mine/mill complex is covered under one provincial labour code, which mandates collective bargaining and the principles of rights arbitration, in lieu of the right to strike during the term of the labour contract. Although each mine has its own collective agreement with its own particularities, an ad hoc form of patterning exists between the companies, promoted through single union representation among the five mines and, in some cases, near common expiry dates for the agreements.

On the other hand, given the differing managerial paradigms in effect at each site, we might expect varying strategies and frequencies in the use of the political apparatuses of production. In brief, we might infer that different production regimes will lead to different patterns of production politics at the workplace. This would contribute to an overall tenor of labour/managerial relations, revealed by the manner in which the political apparatuses of production are deployed by the participants. Aggressive use of disciplinary measures or grievance procedures, for example, eventuating in stacks of grievances and charges, or in problems renewing collective agreements, may be indicative of one type of climate, just as the opposite conditions may be indicative of another. Finally, a specific industrial relations climate may provide either opportunities or disincentives for labour and/or management to realize their objectives in the labour process.

It is in this vein that claims have been put forward on behalf of post-Fordist production relations. Advocates of such paradigms cite the zero sum, adversarial nature of Fordist designs. In an increasingly globalized production matrix, businesses, workers, and even national economies can ill-afford the costs associated with the adversarial production politics that Fordism invited. Among the chief culprits in this diagnosis is the job control unionism that accompanied mature Fordism. Rather than challenging scientific management and its effects in the workplace, 'the job control strategy welded industrial unions and scientific management to one another.'[6] Bureaucratic and rule bound, this form of unionism actually had the effect of disempowering workers, while straitjacketing managerial initiative.[7] As explained by one set of observers, 'In this system of job control unionism, industrial democracy is *reduced* to a particular form of industrial jurisprudence in which work and disciplinary standards are clearly defined and fairly administered and disputes over the application of rules and customs are impartially adjudicated through the grievance procedure [my emphasis].'[8]

Under this system, unions increasingly assumed a regulatory role, ensuring the fair administration of the status quo through seniority provisions, rights clauses, and the grievance machinery of the collective agreement. Grievances in turn, became the property of unions, to be processed by experts (often lawyers) in time-consuming and often costly manoeuvres. Unions and their members were correspondingly denied a broader, more meaningful role in the organization of the workplace and future decision making pertaining to it.[9] Symptomatic of these trends, and perhaps an expression of their highest development according to one observer, is the CWS system of job and wage fixing.[10] Rather than mounting a challenge to the constricting job structures of Taylorism, CWS represented an accommodation of it. In return for accepting the basic parameters of this form of job organization, labour won the right to comanage it, by retaining representation on job evaluation committees. Henceforth, scientific management would be applied in a joint rather than in a unilateral fashion by management and trade unions. For critics of job control unionism, however, retention of the basic postulates of scientific management rendered systems such as CWS an ambiguous gain, at best, for workers. In short, then, a production politics that is premised upon job control unionism both mirrors and actively reproduces the most detrimental qualities of scientific management – worker apathy, disassociation, and disempowerment among workers and their organizations.

New production relations, according to the pundits, hold out the promise of a remedy for the stalemate of Fordism. The key here is the reorganization of work around team principles. The utilization of multifunctional work teams effectively does away with job control unionism. Instead of the one worker/one task form of organization and the conflict that characterizes Fordist production politics, teamwork is supposed to embody the practices of job rotation, flexibility of assignment, and multiskilling. This is reputed to have direct and positive effects on the labour process. More challenging and satisfying work, along with greater levels of responsibility and autonomy, are the often-cited benefits of team approaches.[11] Beyond this, however, a dedication to post-Fordist principles of organization is thought to bestow additional advantages in the conduct of the production politics of the workplace.

In essence, such new production relations are said to entail a trade-off between work rules, which are forgone by the union, and new opportunities for participation by unions and workers in the governance of the workplace. In return for dropping complex job classification systems and narrow lines of demarcation with their corresponding wage rates, labour is offered new venues for involvement in the strategic decision making of the firm. These may range from seats around the corporate board table, to greater involvement through joint union/management steering committees, to direct participation through membership on cross-occupational work teams. More concretely, this may translate into worker participation in facets of the job such as design and layout, training, the setting of work assignments, holiday and overtime scheduling, the adoption of new technologies, and the selection of supervisors and co-workers.[12]

For some, this stands for nothing less than a significant growth in the democratization of the employment relationship, which ultimately will pose more interesting challenges for management than for trade unions.[13] Accompanying this are gains for both labour and management: possibly fewer grievances, a more pleasant, cooperative work environment, and a more agile way of doing work and business.[14] In short, proponents of post-Fordist work relations point to the win/win potential of such arrangements for business and labour.

Critics relay a different scenario. Although workers willingly embrace democratizing initiatives and are often enthusiastic partners in plans that offer such prospects, they are destined to be disappointed by what they receive in the new managerial paradigms. This is the case for several reasons. First, participation schemes are usually framed by management, before being presented to workers as something to 'sign on' to. In other words, the terms of participation including the parameters are, to a large extent, pre-set.[15]

For example, the training needs that are often associated with Employee Involvement or continuous improvement programs are usually defined unilaterally by management in terms of what they think workers need to know. As a result, 'training' may consist mainly of problem-solving exercises dealing with resource allocation, statistical process control, time study methods, or techniques of peer evaluation.[16] Such methods are not, however, agenda free. They embody principles of control over both the self, as well as over others.

Analogously, in the process of work reorganization, jobs are often restructured by engineers and tried out by supervisors before they are staffed with workers. Employees, of course, are welcome to make suggestions to improve jobs that have already been predesigned and laid out for them, but this would be an after-the-fact occurrence. As a result, for some commentators there is less here than is proclaimed; any changes would be at the margins to jobs that had been essentially predesigned to meet corporate requirements.[17]

On top of what may be termed the initiating power of management, the power of the purse also remains a potent factor. That is, managerial control over the expenditure of funds assures ultimate control over the adoption of suggestions that are made. In effect, financial clout lends management a veto power that workers are lacking when they enter into joint involvement schemes. Consequently, the literature is rife with examples of management backing away from specific suggestions and whole involvement plans that challenge established ways of doing things.[18]

Finally, those sceptical of post-Fordist alternatives show concern over the corrosive effects which an abandonment of job control issues may have on unions themselves. Such initiatives as Employee Involvement are often accompanied by a sense of corporate crisis and the threat of job loss. Failing to cooperate, or consenting to go into involvement programs that eventuate in job losses or unfulfilled promises of improvement, may eventually have a negative effect on the union. Adding to the complications, management-structured work teams may provide an alternative source of identity formation for their members, especially if teams are encouraged to compete against one another for available work or production bonuses. As Wells points out, in such instances the team leader may effectively compete with the shop steward for team member loyalty, while the participation plan may substitute itself for the union as a recognized source of empowerment.[19]

When local union officials assume leadership roles in participation campaigns the chances for such confusion are multiplied further. The dualisms that are represented by occupying shop steward positions on behalf of the union and team leadership functions for the company become increasingly difficult to reconcile.[20] As a result, some investigators have come to conclusions that are quite different than those reached by promoters of new work relations. They report upon an inverse relationship between ability to personally influence job conditions and interest in the union.[21] Under some conditions, then, a decline in union efficacy may coincide with the adoption of new structures and practices. This could result from giving up old protections, such as rules and protocols associated with job control, or from the adoption of new parallel structures that threaten to marginalize and displace the union's presence in the workplace.

For the critics of post-Fordism, teamwork and employee involvement programs remain 'a far cry from democratic work organization and autonomous worker decision making.'[22] Instead, they entail an augmentation of managerial power, at the same time as they represent a weakening of labour's presence in the employment relationship. As Wells starkly records, 'The result is less equality between labor and management, less democracy in the workplace.'[23]

Advocates of new work relations, including post-Fordist variants, reach conclusions on issues of control, climate, empowerment, and democratization, that could not be more starkly different from their critics, as the preceding discussion demonstrates. The following sections investigate these issues at PCS, Agrium, Central Canada Potash, and Cameco, first by examining reported incidents of different conflicts at the mine sites. The remainder of the chapter compares actual and desired levels of worker control over various aspects of the employment relationship, degrees of union activity, and the quality of the work environment, more broadly conceived. Under the latter topic, workplace health and safety, job stress, and the important but neglected issue of harassment in the workplace are considered. Analysed separately, and taken together, these features ought to provide a reasonable overview of convergences and variations in production politics under the differing managerial regimes at the five sites of production.

Control and Conflict at the Mines

An immediate glimpse into the industrial relations climates at the five mine/mill sites may be gained by beginning with an inspection of disciplinary practices and organized (that is, recorded and documented) forms of resistance on the part of workers and their organizations. Discipline, of course, is management initiated. It consists of actions brought against individuals singly, or in groups, for alleged infractions of company rules. In spite of detailed codes that prescribe specific offences and appropriate penalties, such as those produced by Key Lake management (see Chapter 3), a certain amount of discretion is always involved in the disciplinary act. Behaviours that are singled out for discipline may vary from company to company and from time to time. Sanctions run the gamut from verbal warnings to disciplinary steps that can lead to suspensions and dismissals. It is therefore instructive to investigate what behaviours are being disciplined by management and what types of sanctions are being invoked.

Simply taking reported cases of personal discipline among the respondents yields a wide range of frequencies between the different sites of production. Four-fifths (82 per cent) of the interviewees at Central Canada Potash reported being on the receiving end of disciplinary actions at one time or another, while 47 per cent of Agrium workers reported being disciplined. The other sites fell at various points in between these limits. When compared with its share of the total sample, workers at CCP are definitely over-represented in their involvement in disciplinary incidents. Although these workers constituted 22 per cent of the sample, they made up 27 per cent of those who had received formal discipline from the employer. Workers at Agrium are to a similar degree underrepresented, again composing 22 per cent of the sample but being only 15 per cent of those who reported being disciplined. Meanwhile, workers at the other sites come quite close to being evenly represented, when respondents receiving discipline are compared with share of the total sample. These figures do not reveal a great deal because they neither represent an aggregation of disciplinary *incidents*, nor do they speak to the severity of the sanctions that were applied, nevertheless the implications are mixed with respect to theoretical expectations. Proponents of post-Fordist work relations might expect lower incidents of discipline as an indication of an improved industrial relations climatic. The fact that fewer workers at Agrium have been subjected to discipline would bear this expectation out. However, this conclusion is mitigated by the comparatively high proportion of employees who have received discipline at Central Canada, the other site that has made the most concerted efforts to move into new work relations. Too much should not be read into the results, however, as a temporal dimension was not built into them. The discipline being captured in these results might have occurred *prior* to the shift into new managerial modes. When the issue of discipline is examined longitudinally, such events were found to be more common at both Agrium and at Central Canada Potash in the earlier years. Disciplinary incidents fell off at Central Canada after 1984, or after the first reform initiative had commenced. At Agrium, the use of such measures tailed off after 1986, but before the movement into team concept and continuous improvement. At the other mines no real pattern is exhibited over time. So, although we might expect memories to dim with the passage of time and, therefore, earlier instances of discipline to be under-reported *vis-à-vis* more recent incidents, this was not the case at either Agrium or CCP. Only at Central Canada, however, was there some prima facie evidence that high rates of discipline decay with the introduction of industrial relations reform.

Taking all reported *incidents* of discipline into account at each site, as opposed to the number of individuals who have been disciplined, reveals an interesting pattern. First, as can be seen in Table 6.1, the highest overall rates of sanctioning were recorded at Allan and Central Canada, respectively, which each accounted for one-quarter of the total number of cases of administered discipline. With the exception of CCP, the most common causes of discipline being invoked were absence without permission or lateness. These were followed at all of the mines by being held responsible for unacceptable work, which might include failure to complete a job in a thorough manner, working in such a way as to be held accountable for an accident, or other improper equipment utilization. Refusal to work figured into the results for Central Canada, but this was owing to the shutdown of the operation in 1981 by a complete and effective two day wildcat strike directed against the firing of a worker. Many employees received discipline for this one episode.

At all mines only a fraction of the disciplinary charges were contested. This could reflect the nature of the incidents and at some mines there is a statistically significant relationship between the action being disciplined and whether or not a grievance is filed.[24] For example, at Agrium, 63 per cent of discipline cases involved absence or lateness, the highest figure for this type of infraction. Overall, only 6 per cent of total disciplinary events at Agrium were appealed to the grievance system and none of the cases around absence have been contested. A willingness to counter disciplinary acts may also be indicative of the local union's stance towards management. Although absence and lateness were also the major factors behind discipline at Key Lake and at Allan, proportionately greater numbers of grievances were filed on this point (24 and 20 per cent, respectively, of all cases involving disciplinary actions against absence). At Key Lake this may signify a willingness on the union's part to try and mitigate

TABLE 6.1
Disciplinary issues

Issue		Total reported disciplinary incidents					Row totals
		Allan	Central Canada	Cory	Cameco	Agrium	
Refusal to	N	7	40	6	6	3	62
work	%	5.4	31.0	8.1	6.3	3.5	12.1
Late/AWOL	N	72	34	30	45	54	235
	%	55.4	26.4	40.5	47.4	62.8	45.7
Attitude	N	5	12	14	14	4	49
	%	3.8	9.3	18.9	14.7	4.7	9.5
Unacceptable	N	37	38	21	24	22	142
work	%	28.5	29.5	28.4	25.3	25.6	27.6
Other	N	9	5	3	6	3	26
	%	6.9	3.9	4.1	6.3	3.5	5.1
Column	N	130	129	74	95	86	514
totals	%	25.3	25.1	14.4	18.5	16.7	100.0

Cramer's V = .21; p = .000.

some of the results of radically spatialized industrial relations and their effects on workers. In other words, absences may be more easily, or willingly, problematized and contested in the context of such relations.

An adage that is often said to characterize production politics in Canada, the 'obey now, grieve later' syndrome, is also in effect at these workplaces.[25] This rule of thumb is not quite as trite as it may first seem. It is an acknowledgment of management's right to manage, in the first instance, without formal challenge. Instead of contesting this right through job actions aimed at reversing unilateral dictums, workers have the right to file grievances through their unions, to the effect that their rights as laid out in the collective agreement have been violated by managerial actions. Grievance proceedings thus become an important *formal* indicator of resistance, although by no means the only, or even the most important sign. In short, where the right to strike is severely constricted, as it is in Canada, rights arbitration through the grievance system takes on an added significance in production politics.

Table 6.2 presents data on the proportion of workers interviewed at each site, who, at one time or another, have filed one or more grievances on their own behalf against managerial actions that were deemed detrimental. The variation, ranging between Agrium at the low end and Allan at the high is impressive and significant. The two PCS mines, Allan and Cory, have the highest proportions of workers who have taken action along this line (61 and 55 per cent of their

TABLE 6.2
Grievances filed

Respondents filing grievances		Mine					Row totals
		Allan	Central Canada	Cory	Cameco	Agrium	
Yes	N	41	34	28	27	24	154
	%	61.2	45.9	54.9	40.3	33.3	46.5
No	N	26	40	23	40	48	177
	%	38.8	54.1	45.1	59.7	66.7	53.5
Column totals	N	67	74	51	67	72	331
	%	20.2	22.4	15.4	20.2	21.8	100.0

Cramer's $V = .20$; $p = .009$.

workforce samples, respectively). Both of these sites are relatively over-represented in this category, accounting for a higher proportion of grievance filers than for the sampled labour force as a whole. At Allan, just over 61 per cent of the respondents indicated filing at least one grievance on their own behalf.[26] Although the Allan sample only accounted for 20 per cent of the total, 27 per cent of the grievance filers worked at this mine. By the same token, the Key Lake operation and Agrium are relatively under-represented by grievance filers; the proportion of workers at these mines who have brought such actions against management are lower than their share of the total workforce sample. The proportion of the sample who had filed at least one grievance at Agrium is roughly half of that for Allan, while at Cameco, 40 per cent of the sample had submitted at least one grievance.

The pattern is clear, but inferences derived from it are less so. In one regard, these results could add support to post-Fordist contentions. The most traditional sites have the highest proportion of grievance filers *in situ* and account for the highest proportions of grievances filed overall (Table 6.3). By these indicators, the most conventional firms also have the most contentious industrial relations climate. The most experimental firm, Agrium, has the lowest frequencies and proportions of employees who lodge grievances and the lowest absolute number of reported grievances among respondents. On the other hand, however, less grievance activity could imply less local union vitality, an interpretation that critics suggest is entirely consonant with post-Fordist production politics. Additionally, as in the discussion of discipline, the issue of when the grievances were being lodged is important for any conclusions that may be drawn from this data.

The issues involved in grievance proceedings further shed light on the production politics at each mine. At Allan, for example, as Table 6.3 illustrates,

TABLE 6.3
Grievance issues

Issue		Total reported grievances					Row totals
		Allan	Central Canada	Cory	Cameco	Agrium	
Insubordination	N	0	0	3	3	2	8
	%	0	0	5.8	5.6	6.1	2.9
Unacceptable work	N	3	5	3	11	1	23
	%	3.4	10.4	5.8	20.4	3.0	8.4
Late	N	4	0	2	8	0	14
	%	4.6	0	3.8	14.8	0	5.1
Overtime/holiday	N	16	7	17	10	12	62
	%	18.4	14.6	32.7	18.5	36.4	22.6
Seniority	N	23	20	10	5	8	66
	%	26.4	41.7	19.2	9.3	24.2	24.1
Out of scope	N	14	4	6	5	3	32
	%	16.1	8.3	11.5	9.3	9.1	11.7
Pay dispute	N	5	6	0	0	0	11
	%	5.7	12.5	0	0	0	4.0
Unjust discipline	N	12	0	0	0	0	12
	%	13.8	0	0	0	0	4.4
Harassment	N	2	0	0	0	0	2
	%	2.3	0	0	0	0	.7
Other	N	8	6	11	12	7	44
	%	9.2	12.5	21.2	22.2	21.2	16.1
Column totals	N	87	48	52	54	33	274
	%	31.8	17.5	19.0	19.7	12.0	100.0

Cramer's V = .31; p = .000.

grievance activity was dominated by questions pertaining to seniority (26 per cent of reported grievances) and supervisors doing out-of-scope work (16 per cent). This pattern is obviously related to the regime of seasonal employment that was introduced at this division in the late 1980s. In the same vein, grievance activity at PCS-Cory was dominated by the distribution of overtime work and the important role that this now plays in the mine's operation, with a radically scaled-down workforce. Seniority issues also dominated at Central Canada Potash, where they constituted 42 per cent of the reported disputes. This may be attributed to the combined effects of retaining a CWS system of progression/ bumping and soft markets in the industry that produce fears of future lay-off activity. Grievances pertaining to seniority issues are in fact proportionately over-represented at Central Canada and at Allan. Such disputes constituted a quarter (24 per cent) of the total grievances that were recorded, for which these two sites account for practically two-thirds of this total. This would indicate an active union role in policing the integrity of the CWS charter at the two sites. The same point can be made with respect to the importance of controlling out-of-scope work at Allan. Given the lay-off situation at this mine, supervisors caught doing hourly work is a point of much resentment. Consequently, the bulk of such disputes (44 per cent) were found at this one mine alone.

Overtime and holiday arrangements were the most frequent source of dispute at Agrium, constituting over a third (36 per cent) of the total grievances filed at this mine. More interesting were the presence of significant numbers of grievances pertaining to seniority issues at this mine. Although CWS does not exist here, and the occupational grid is exceedingly flat, this is no guarantee that the distribution of work and opportunity through internal labour markets will not be a contentious point. Indeed, one-quarter of the grievances at Agrium, the second highest proportion of disputes at the mine, were still around seniority issues and the role that they play in determining work assignments and training opportunities. De-emphasizing job control, does not, apparently, negate the factors that gave rise to it, as the Agrium case shows. In fact, on this one issue, the company workforce sample mirrored almost perfectly the industry sample as a whole in its propensity to submit grievances on this issue.

At Key Lake a predominant pattern is less in evidence. That the most frequent cause of grievances at this site was related to charges of unacceptable work standards (20.4 per cent of total reported grievances) may be more related to the hazardous nature of the product being handled than to anything else. This one issue was closely followed by scheduling disputes (overtime and holidays) and lateness, both reflections of the impact that radically spatialized work relations have on workers' lives.

Grievance activity, as reported to us, has marginally fallen off at Agrium since 1989, when the Proudfoot initiative into Taylorization proper was displaced by the continuous improvement project. Elsewhere, the opposite is more common. Allan has witnessed a definite upswing in the frequency with which grievances are reported since 1987, which coincides with the regularization of lay-off activity by the company. Cory division has followed suit, with a marked increase in grievance frequency since the early 1990s, presumably as a result of the demands that have been placed on workers in the advent of adopting radically marketized work relations and the ethic of doing as much, or more, with less. Reported grievances are also more common at Central Canada Potash in the last few years in spite of forays into new production relations, while at Cameco a definitive trend is not present.

To momentarily recapitulate, one point seems noteworthy. Levels of *documented* conflict, as measured either by proportions of workforces involved at each mine, or by total recorded incidents of managerially instigated discipline and grievance activity, vary in a significant fashion from site to site. Regardless of the measure (discipline or grievance initiation), or the unit of analysis (individual respondents or aggregate disciplinary/grievance activity), Agrium reported the lowest levels of such conflict. For example, 12 per cent of all recorded grievances were registered at this site, compared with 19 per cent at the much smaller Cory operation and 32 per cent at PCS-Allan (column totals, Table 6.3). At the other end, Allan exhibited higher proportions of workers who file grievances, higher overall rates of grievance activity and high (although not the highest) proportions of workers who have had disciplinary actions taken against them. The other study sites fall in between these limiting cases. In short, there was less sanctioning activity by management and less grievance filing by workers at the site that most clearly embodies post-Fordist employment relations. This finding though, merely begs the next question, why is this the case?

As suggested previously, this result could be compatible with two opposing explanations. Lower levels of documented conflict could be the outcome of work relations that have removed irritants to the employment relationship. Or, it could herald from forms of work organization that diminish union power in the workplace. In order to scrutinize this important issue more carefully, and as a first approximation, respondents were queried in detail about their participation in the life of the local unions. Participation could range from voting in periodic leadership elections and contract ratification votes at the low end, to serving in an official capacity for the union at the upper end. On the basis of the number of activities partaken, respondents were classified as either being 'minimalists' or 'activists.' The former group consisted of the more passive members; those who would vote and occasionally attend monthly meetings. 'Activists' consisted of the volunteers who ran or campaigned for office and who agreed to participate in the joint committee work of the union (grievance handling, collective bargaining, and occupational health and safety committees), as well as taking the necessary training to be able to do so. Fewer volunteers and more minimalists could be taken as a danger signal for the union, signifying an absence of support on the part of the membership. Alternatively, greater numbers of volunteers could be taken as an indicator of energy within the local.

In fact, there is not a great deal of variation from one mine to the next. Key Lake stands out as having a high rate of participation, but this is easily accounted for by virtue of the rotational week in, week out labour system at the mine. Having, in essence, two different labour forces, week one and week two, necessitates participation by greater numbers of workers, just to keep the union functioning. Apart from Key Lake, 'activists' constituted between 20 per cent (Central Canada Potash) and 35 per cent (PCS-Cory) of the sampled memberships. In this respect, Agrium is not exceptional, one way or the other. With 'activists' comprising 26.4 per cent of its sample, this mine was just slightly under-represented in the total sample of 'activists' along with Allan and Central Canada Potash. Overall, there was no statistically significant variation between the mines, and the association between level of membership activity in the union and worksite was not a powerful one. In other words, nothing suggests that participation levels in the local union have decayed as a result of new managerial initiatives at Agrium. Militancy is another question, which may or may not have been affected, but this is a different question from that which is most relevant here. Changing paradigms at Agrium have not resulted in any apparent atrophying of union participation as compared with the other mines. As for militancy, it is worth recalling that the only collective job actions to occur during the course of this study took place at Agrium, in part as a result of impatience with the continuous improvement initiative and its perceived failure to deliver what had been held out by management.

The analysis of conflict, to this point, has dealt with documented instances, for which there are physical records of discrete events. This by no means exhausts the topic, however. A good deal of resistance goes on beneath the surface of official narratives, unrecorded and perhaps even unnoticed for much of the time.[27] This does not make it any less real than the strikes that dominate local headlines, or the grievances that occupy much of the union's official business. One form of conflict or resistance, which is consequential but seldom receives much attention, is sabotage. For the purposes of this analysis, sabotage is defined as conscious, wilful behaviour that results in loss or damage. This is to be distinguished from careless, sloppy, or what miners term 'rammy' conduct that may characterize working practices. To be a poor driver, for example is one condition. To handle company vehicles in a way that one would not dream of driving one's own car is another, quite different state of affairs. The first implies an absence of skill or competence, the second an *intentional suspension* of those qualities. Although such acts are often individual in nature, the work group may connive in them through offering various types of support, ranging from approving jokes about specific incidents, to covert aid in masking perpetrators. Joking about equipment 'breakdowns' after lay-off notices had been received was standard fare and an example of tacit approval on the part of work groups at one of the mines studied. To take a more dramatic example, at one mine, a new vehicle was rolled in an area where this would be a very difficult accomplishment, and yet no one could recall who the driver was! Both of these examples would fit the definition of sabotage offered here.

On this topic, our approach approximates in some measure Edwards and Scullion, who define sabotage as 'deliberate behaviour leading to the destruction of, or damage to, the company's property; this includes deliberate poor work as well as the destruction of existing work.'[28] In turn, Edwards and Scullion follow the lead of Taylor and Walton who tie sabotage up with 'rule breaking which takes the form of conscious action or inaction directed towards the mutilation or destruction of the work environment.'[29] The further distinction that these authors introduce between utilitarian sabotage and broader forms that are aimed at exerting control over the labour process is less helpful. The former refers to acts that are intended to optimize worker earnings. Essentially these motivations apply to piecework situations, where workers let quality slide in favour of optimizing quantity and hence, income. As Edwards and Scullion recognize though, a certain level of such activity may in fact be in management's interests as well. Thus, there may be a grey line between sabotage in this sense of the term, and simply taking production-maximizing shortcuts. This would have to be determined empirically. In any event, since piece rates or simple production bonus systems were not in operation at any of the sites in this study, this type of dynamic is not especially relevant.

The second category of sabotage, which Edwards and Scullion consider, is equated with the withdrawal of cooperation. Again, however, this operationalization slides too easily into other phenomena. The withdrawal of cooperation, for example, may entail working to rule. Here the dynamic may be considered to be the mirror opposite of what is normally considered sabotage, with extra special care being taken to ensure rule compliance and quality (100 per cent quality control!) albeit at the expense of meeting production targets.

With these points in mind, workers were asked a number of questions pertaining to the witnessing of what they took to be deliberate acts of negligence on the job. The questions were gradated, commencing with observations about deliberately running equipment too fast and leading up to acts of vandalism. The last item is the most definitive. On this point, one-third of the workers who were interviewed at Agrium reported witnessing intentional acts of equipment breakage, compared with 34 per cent at Central Canada, 22 per cent at Allan, 18 per cent at Key Lake, and negligible reports at Cory. Interestingly, in this category of conflict, Agrium and Central Canada were equally over-represented and in a statistically robust fashion.[30] Higher proportions of workers at these mines reported on vandalism than was the case overall. Intriguingly, almost identical amounts of deliberate mishandling of equipment were reported at four out of the five work sites, again including Agrium, Allan, CCP, and Key Lake, where between 55.6 and 56.8 per cent of the respondents indicated observing this type of action (Table 6.4). Curiously, the proportion of respondents reporting on all types of sabotage was much lower at PCS-Cory than elsewhere. That this is an older workforce that

TABLE 6.4
Reported levels of different types of sabotage

	Respondents reporting on activity (%)				
Type	Allan	Central Canada	Cory	Cameco	Agrium
Running machines too fast*	37.3	31.5	8.0	32.8	23.6
Improper handling of equipment	56.7	56.8	17.6	56.7	55.6
Vandalism*	22.4	34.2	2.0	17.9	33.3
Deliberate waste*	73.1	74.0	29.4	53.7	75.0

*Significance ≤ .02.

has had its expectations of job security shattered by a radical downsizing may have something to do with these responses, although there is nothing in our data which would permit us to come to a definitive conclusion on this point.

Overall then, although the forms of production politics do get altered under the auspices of different managerial regimes, it is hardly the case that they disappear, or that conflict is banished from them. Worksites, such as PCS-Allan where lay-offs and seasonal employment have been substituted for employment security, are rife with resentment over declining living standards and the daily realities of supervisors who continue to earn full pay cheques during lay-offs. Disciplinary and especially grievance records are a testimony to this, while at the same time workers are more reluctant to take on strike action out of fear of losing more work time.

At other mines, these are simply not issues. Furthermore conflicts around new initiatives such as those displayed at Agrium and Central Canada are often not grievable issues. But the conditions for more collective forms of job action may be more propitious and have been propelled forward by disappointments associated with the inability of new workplace initiatives to measure up to expectations. Hence Agrium, despite the hopes placed on non-adversarial production politics has recently had two strikes in three years, the first lasting for one day and the second, discussed more fully in the next chapter, accounting for four months of lost wages and production time. Similarly, the recent unilateral withdrawal of the lucrative performance-sharing plan at Central Canada by its new owners may be expected to have telling collective implications.

Finally, at each site, with the partial exception of Key Lake, the actual labour process remains similar, as do opportunities for sabotage as a form of 'getting some of one's own back at the company.' New production relations notwithstanding, they appear not to have made much of a dent in this form of worker resistance. PCS-Cory is the only outlying case when it comes to reported sabotage.

The question as to whether new production relations mean better industrial relations needs qualification – better for whom? Thus far, the impact of differing managerial paradigms on indicators of conflict at the worksite have been examined. Although a reduction in grievances, disciplinary action, sabotage, or strike activity may or may not entail benefits for workers, clearly it represents a considerable achievement for management. A reduction in conflict is viewed as a desideratum by managers as well as the keepers of the national income accounts. Generally, it is equated with a higher quality of work effort as well as superior productivity. But does it also entail better work environments for employees?

Production Politics and the Work Environment

The notion of work environment invoked here is intentionally broad. It takes in health and safety issues, but goes beyond them in such a way as to import related concerns back into the underlying principles that subsume our notions of health and safety. For example, I take it as an axiom that a safe working environment is one that is free of personal harassment from supervisors and co-workers, as well as one that is free from exposure to dust, dangerous toxins, or bad ground. In the same fashion, I would understand a healthy workplace to be one that enhances the direct participants' control over the work they perform in such a manner as to reduce the job stress that may be associated with an absence of control. To wit, worker participation ought then to encompass not only adaptive issues of corporate concern, but the very framing of those areas that are open for joint input.

Owing to the importance of the issue of control over decision making in the various arenas of working life and its salience for any discussion of the quality of the work environment, we begin with it. In the workplace survey, respondents at each worksite were queried as to whether they had any real involvement in the making of decisions on a battery of issues, ranging from job training and work layout to decisions that have a direct impact upon management, such as input into the selection of supervisors and the setting of managerial salary levels (Table 6.5). In addition to this, for those mines and for those realms where labour was denied an effective role, workers were asked if they would like to have greater input. To avoid spurious responses in this section, it was made clear that participation would involve both time and effort, such as regular attendance at meetings and a commitment to take on joint committee work with management. The results, by mine, are contained in Table 6.5 and Table 6.6.

First, with respect to the existent situation, the most striking feature is the overall modesty of any real variation between the worksites. Not only is there an absence of significant variation on most of the participation indicators between the mines, but there is no overall pattern either. In some cases the differences that exist are in the direction that might be predicted by exponents of post-Fordism, as when worker participation is greater under the more reform-oriented managements. But in other instances the findings show just the opposite; greater levels of reported participation under more traditionally inspired managements.

To take some examples, larger proportions of the samples at Agrium and Central Canada Potash reported involvement in deciding how job training would be carried out, in having input into work assignments and quality control issues, and in determining production targets. These results are as the advocates of new work relations would predict, but they immediately need to be qualified by two additional points. Within most of the relevant domains, including training, setting work assignments and production targets, and taking a lead in quality control, it is still minorities who indicate participation in the particular activity. For instance, at Central Canada Potash and at Agrium, 36.5 per cent and 32 per cent, respectively, stated that they had input into the way job training was done. The proportions are lower for other areas of control, such as the setting of production targets and work assignments. Second, the reported results from these mines do not differ in a statistically significant fashion from the other sites. In the case of job training, for example, participation at Central Canada and Agrium was only marginally greater than at Cory and somewhat greater than at Allan or Cameco.

Where significant variation does exist, the results are indeed more mixed. Workers at Central Canada and Allan reported participation in higher proportions over the establishment of work pace and in the adoption of new technologies. Parenthetically, at both of these sites, CWS is staunchly defended by the local union. On more overtly political issues such as having a say in the promotion of co-workers or conducting appraisals of supervisors, workers at Agrium and Central Canada are clearly advantaged, although not so much because of the great extent of these practices at these mines, as because of their total or near complete absence at the other sites. Finally, the omission of a co-determining role on any of the more macroscopic issues of managing the workplace is notable. Participation and empowerment apparently do not extend to such matters as salary and dividend payouts, or future placement of investment funds at any of the mines. They only infrequently extend at some sites to such related issues as hiring on new employees or selecting supervisors, and then on an experimental, ad hoc basis.

TABLE 6.5
Reported levels of worker involvement

Involvement	Sample indicating involvement (%)				
	Allan	Central Canada	Cory	Cameco	Agrium
Job training	25.4	36.5	29.4	19.4	31.9
Setting job assignments	13.4	24.3	13.7	22.4	27.8
Job layout	50.7	56.8	39.2	40.3	55.6
Quality control	29.9	43.2	37.3	30.3	40.3
Setting work pace*	86.6	89.2	68.6	74.6	72.2
Setting production targets	10.4	16.2	3.9	9.1	18.1
Adopting new machinery*	40.3	56.2	15.7	28.4	34.7
Formulating safety rules	41.8	47.3	51.0	47.8	34.7
Hiring new workers	0.0	2.7	3.9	1.5	0.0
Promotion of co-workers*	3.0	4.1	0.0	1.5	15.3
Discipline of co-workers	0.0	5.4	3.9	7.5	5.6
Selection of supervisors	0.0	1.4	3.9	1.5	5.6
Yearly evaluation of supervisors*	0.0	13.5	0.0	0.0	5.6
Setting management salaries	0.0	1.4	0.0	0.0	0.0
Setting stockholder dividends	0.0	0.0	2.0	0.0	0.0
Setting investment policy	0.0	0.0	2.0	1.5	0.0

*Significance $\leq$ .05.

In other words, as Table 6.5 graphically illustrates, managerial rights are still an exclusive preserve. Indeed, it's as if an invisible line had been drawn through the table. On the strictly labour process issues that stretch from involvement in job training to input into the formulation of safety rules, workers at all of the mines do participate, sometimes more and sometimes less, at the sites that have introduced formal mechanisms for involvement. After this point, however, inclusion is rare to non-existent. Only 15 per cent of Agrium workers indicated any involvement in promotion decisions, while 13.5 per cent of the Central Canada interviewees had participated in supervisory evaluations. And these were the best case exemplars.

It has been suggested that the patterns in Table 6.5 may reflect worker preferences. In other words, despite the assumptions of socialist theorists, workers may not place as high a priority on involvement in workplace governance as their erstwhile spokespersons do. Moving away from the labour process with its immediate impact on workers, enthusiasm for co-determination may wane as the decision-making environment becomes more foreign to workers.[31] Workers may feel neither interested in nor competent in those areas of decision making that have traditionally been reserved for management under the doctrine of residual rights. Encouragingly, little evidence was found in support of such generalizations. Even when greater influence was tied to more demanding time commitments for the requisite training and subsequent committee work, impressive numbers of workers, who do not now consider that they have a voice in important control issues, indicated a willingness to partake in corporate decision-making responsibilities. Majorities at all mines, with the exception of Cory, were interested in having input into the selection and yearly evaluations of their supervisors, an initiative that would surely go some way in curtailing day-to-day unilateralism in these workplaces. Majorities at Allan and Agrium also wished to exercise influence over managerial salaries, as did healthy minorities at Central Canada, Cory, and Cameco. Even in the more technical areas, such as corporate investment and dividend policies, between one-fifth and one-third of the samples, depending upon the issue and the mine, stated a preference for greater involvement. Tellingly, on these issues of macrocontrol (salaries, investments, and dividends), there was no significant variation between sites. This absence of variation in the proportions wanting greater levels of participation in the more global areas of concern, in all likelihood, reflects the very real situation of total exclusion from these same policy-making areas, which is common across all of the mines.

Significant differences in the numbers who are willing to assume more responsibilities for comanagement are registered in the areas of job training, the setting of work assignments, job layout/design, hiring, disciplining, and selection/evaluation of supervisors. With all but one of these issues, proportionately more workers at Allan express an interest in extending the frontiers of control. This may be related to actual conditions at the mine. As noted in the previous section, Allan had the highest rates of both reported disciplinary incidents and submitted grievances. Although the workforce at this mine constituted 20 per cent of the total sample, it accounted for 25 per cent of the total caseload in disciplinary actions and an even higher 32 per cent of the total reported grievances. Coinciding with these trends is the relative sense of disempowerment that has been brought on by the ongoing use of indefinite lay-offs at the mine. Small wonder then, that a proportionately larger share of the workforce at Allan would show an interest in joining in on such matters as the selection and evaluation of supervisors, along with attaining a voice in promotions and directly determining whether discipline is justifiable in any given instance.

Equally unremarkable is the result indicating that more workers at Key Lake than elsewhere would be interested in taking part in hiring decisions. Indeed, Cameco is the one company that may possibly remain in a hiring mode over the foreseeable future. On top of this, its operations are carried out in regions of high under- and unemployment, where employment equity considerations are supposed to govern. In the north, there is a feeling that more of the jobs in the uranium industry should be going to local residents. As well there is a perception that the industry should make training and advancement opportunities available for the aboriginal workers, which are not currently evident (see Chapter 2). With this as background, the higher proportion of workers who would like to be more involved in hiring decisions is readily comprehensible. At both sites, Allan and Cameco, existent conditions help generate highly rational responses on the part of the workforces.

As indicated in Table 6.6, greater involvement in the establishment of safety measures and rules ranks across the sites as one of the issues in which high proportions of workers stated they would be most willing to further their involvement. This underscores the general importance which health and safety issues have in this industry. If industrial systems generally are unnatural environments, then mining in particular presents the worker with a most alien set of conditions. Potash production, as detailed in the last chapter, entails the use of mammoth equipment in underground settings. Occasionally, this is supplemented with the use of explosives.[32] In addition to the sheer scale of the technology that is currently used in underground operations, ground conditions can vary considerably from one section of a mine to another. 'Bad back' can result in large sections of razor sharp rock shearing off of walls without warning, and more than one miner has been badly injured in such incidents.

Throughout the interviews, stories were recalled of accidents and near

TABLE 6.6
Willingness to become involved in joint decision making

Decision-making issues	Sample indicating willingness to be involved (%)				
	Allan	Central Canada	Cory	Cameco	Agrium
Job training*	86.0	61.7	50.0	72.2	67.3
Setting job assignments*	52.6	33.9	27.3	40.4	26.9
Job layout	78.8	40.6	48.4	47.5	50.0
Quality control	25.5	14.3	6.3	23.9	20.9
Setting work pace	77.8	50.0	37.5	35.3	45.0
Setting production targets	27.1	21.0	12.5	20.0	27.1
Adopting new machinery	82.5	71.9	69.8	64.6	76.7
Formulating safety rules	74.4	59.0	52.0	60.0	66.0
Hiring new workers*	22.4	22.2	20.4	50.0	27.8
Promotion of co-workers	42.4	31.0	25.5	37.9	34.4
Discipline of co-workers*	35.8	15.7	16.3	30.6	5.9
Selection of supervisors*	70.1	56.2	42.9	56.1	61.8
Yearly evaluation of supervisors*	83.6	81.3	54.9	65.7	72.1
Setting management salaries	53.7	38.4	39.2	40.3	52.8
Setting stockholder dividends	27.3	18.9	20.2	24.2	18.1
Setting investment policy	30.3	23.0	22.0	27.3	22.2

*Significance ≤ .05.

misses. One death occurred at Central Canada Potash shortly before we started interviewing there. An operator became ensnared in the tracks of a caterpillar loader when it unexpectedly started moving backwards off of the tailings pile he had been working on, just as he was attempting to dismount the vehicle. This brought the number of deaths at this mine to eight since first opening in 1969. Prior to this accident, in 1991 the local union president at the Cory mine, a journey electrician, was instantly electrocuted while working on an underground transformer. A mining operator, whom we had earlier interviewed at Allan, came close to losing his life when he was hit by a five-ton piece of rock that had sheared away from the roof just as he was climbing down from the miner. In this near-fatal incident, the worker was trapped between the wall and the side of the miner by the fallen rock, waiting with one punctured lung, several broken ribs, and a fractured skull, while fellow crew members manually removed the fallen slab.

Perhaps most hauntingly of all, just one day after I had completed the first draft of this section of the chapter, two workers were killed and a third badly burned at the Cory mill. Working in the mill atop a large holding tank, filled with a 95 degrees centigrade brine solution, the workers were dumped into the scalding solution below when the cover unexpectedly gave way. In this preventable accident, young contract workers had been brought into the mill to do repair work under conditions that the plant's own, unionized workforce would not have agreed to. As this tragedy attests, mill operations can also pose deadly risks. In a hot and noisy environment, maintenance and upkeep work in the mills often entails working at considerable heights off of gridiron floors, or on top of large holding vessels. At Key Lake, there is also the added consideration of exposure levels to radioactive material.

Even work in an open-pit environment is not without its dangers. At Key Lake, a worker accidently backed the haulage truck that he was operating out over a bench in the pit. Miraculously, the worker survived the thirty-foot fall in the truck to the next bench level, although the vehicle was a write-off.

Until recently, official data on work accidents has been notoriously misleading. The main collecting agencies were the various provincial workman's compensation boards. However, owing to the fact that compensation premiums, as well as corporate reputations are tied to safety records, there is considerable pressure to under-report accidents to the boards.[33] Only time-loss mishaps, that is, injuries that result in taking time off of the job and thereby result in a claim, are entered into this data source. Given this practice, there is pressure to reassign workers who have been hurt to light 'housekeeping' duties, in order to keep the incident off of the company's claims record. In this manner, mishaps in the past may have disappeared from the official record.

TABLE 6.7
Work injuries

Reported injury		Mine site					Row totals
		Allan	Central Canada	Cory	Cameco	Agrium	
Yes	N	58	49	42	42	53	244
	%	86.6	66.2	82.4	62.7	73.6	73.7
No	N	9	25	9	25	19	87
	%	13.4	33.8	17.6	37.3	26.4	26.3
Column total	N	67	74	51	67	72	331
	%	20.2	22.4	15.4	20.2	21.8	100.0

Cramer's V = .21; p = .007.

In addition to those WCB records that are filed with the mine inspectorate, there is now also a requirement for mining companies to file notice of serious bodily injuries and dangerous site occurrences with the provincial mining inspector within twenty-four hours. Although this may improve upon the quality of the official statistics, there is no indication that the separate files from the mine inspectorate and the WCB have been merged, or that they are readily available to researchers. Thus, to avoid some of the problems with the use of 'official' statistics, workers in the study were asked a number of questions pertaining to their own accident records at the mines. This began with inquiring as to whether the respondent had ever incurred a serious injury. 'Serious' was defined as an incident that went beyond the cuts, bruises, and nicks that can often be a part of manual labour.

Given our critique of the inadequacies of the existing data bases on this subject, the results still came as a surprise. At the two PCS divisions, over four-fifths of the sample had been injured on at least one occasion (Table 6.7). Although both of these mines are slightly over-represented in reported injuries, the record elsewhere is almost as disconcerting. Two-thirds of the Central Canada sample and nearly three-quarters of the Agrium sample had also suffered on the job injury. Amongst these mishaps, the most common in the industry appear to be back injuries. They account for fully one-third of the injuries that were registered in our survey.

Lest it be supposed that too much is made of these figures, we can switch to the time-loss approach, as is done in Table 6.8. It divides the experiences recorded in the previous table up in accordance with whether or not the individual has undergone a time-loss injury. Again, the figures are disturbing. At Key

TABLE 6.8
Time-loss work injuries

Time-loss injury		Allan	Central Canada	Cory	Cameco	Agrium	Row totals
			Mine site				
Yes	N	38	26	20	29	23	136
	%	65.5	53.1	47.6	70.7	43.4	56.0
No	N	20	23	22	12	30	107
	%	34.5	46.9	52.4	29.3	56.6	44.0
Column total	N	58	49	42	41	53	243
	%	23.9	20.2	17.3	16.9	21.8	100.0

Cramer's $V = .21$; $p = .03$.

Lake, 70 per cent of the workers who reported an injury required time off work. At the potash mines, the percentage of the workforce samples requiring time away from work after suffering an injury ranged from 43 per cent at Agrium, to 66 per cent at Allan. Workers at both Allan and Key Lake are over-represented in the time-loss group, as against the accident category as a whole.

Workers in the sample were also asked whether they suffered from any other forms of ill health, apart from the injuries described above. There was little variation between the mines. One-third of the total sample provided evidence of more general health problems, while the proportion at each mine ranged between 24 per cent and 39 per cent of the respective samples. More interesting, perhaps, is that a large proportion of these groups associated such problems with their work. Large majorities at four of the five mines (64 to 76 per cent) drew a connection between the general health problems they were experiencing and working in a mine or a mill.[34] Once again, back ailments led the way in extraworkplace health concerns. Lung problems (asthma, et cetera) and arthritic complaints were also notable by virtue of their inclusion within the category of general health problems that were experienced by the sample. Clearly, the connection between these ailments and heavy equipment operation in the hot dusty environments of the mines and the hot, humid conditions of the mills has not gone unnoticed by the subjects of these labour processes.

At least some mine accidents could have been foreseen and are related to the performance of unsafe work. In the survey, we were interested in ascertaining how common an experience the request to do unsafe work was. Such demands could entail a variety of situations. Entering into an unsecured area, using unsafe equipment to do a job, being asked to do work that one is unfamiliar with or ill trained for, or being asked to do a job using improper procedures so as to get it done faster or more economically, are all examples of possible scenarios. By the same token, workers may respond to such instructions in different ways. Under law, *individual* workers may refuse a work order that they consider to be hazardous. If pushed, this may lead to an investigation of the condition by the mine inspectorate. Then again, in an effort to avoid an inquiry, supervisors may simply postpone such jobs until the next shift in an attempt to find a more compliant worker who is willing to go along with a supervisor's instructions. How workers respond to requests to do work that they consider to be dangerous is also a crucial element in the production politics at each site.

As with industrial accidents, instructions to perform work that participants in the labour process deem dangerous is a common experience. In fact, at the four potash sites, there was less than a five percentage point spread in the proportions of the samples who had received orders to undertake work which they considered risky. At Agrium, 61 per cent of the sample reported on receiving instructions to do a job which the incumbent considered unsafe, while at Allan the figure rose to 66 per cent. Key Lake is something of an anomaly, with 48 per cent of the sample stating that they had been asked to do unsafe work. This is probably accounted for by the open-cast nature of the mining techniques at this site. Even with the Key Lake operation as part of the comparison though, there are no statistically significant differences in the proportions of workers who are asked to undertake questionable jobs at each mine. In short, it is a common experience for large proportions of workers at each operation to be asked to perform jobs which they consider risky.

In a majority of cases, instructions to do work which is considered potentially dangerous entails working in improperly secured areas. Inadequate scaling, bolting, or ventilation are examples of such underground occurrences, while working in a dirty environment may pose risks of falls or other injuries in the mill. At each site, more than half of the reported incidents to engage in a job that was considered dangerous involved working in an unsafe area. Orders of this type constituted 57 per cent of the total number of unsafe work requests at Allan and 81 per cent at Agrium. Use of unsafe procedures was also a commonly cited situation across the mines. This often entailed taking shortcuts in doing a job, such as not fully locking out electrical equipment when doing routine maintenance work on it. This latter was one of the most usual concrete examples provided of the use of unsafe working procedures.

Responses by workers to requests to do jobs that they consider to be unsafe is an integral aspect of the production politics or political culture of any workplace. Unlike the proportions of workforces that were affected by requests to

TABLE 6.9
Responses to unsafe work requests

Response		Allan	Central Canada	Cory	Cameco	Agrium	Row totals
			Mine site				
Went along	N	23	47	16	25	64	175
	%	20.0	38.8	21.1	33.8	42.4	32.6
Objected, then did it	N	21	6	9	20	31	87
	%	18.3	5.0	11.8	27.0	20.5	16.2
Refused to do it	N	55	61	29	27	51	223
	%	47.8	50.4	38.2	36.5	33.8	41.5
Refused, referred to union	N	16	7	22	2	5	52
	%	13.9	5.8	28.9	2.7	3.3	9.7
Column totals	N	115	121	76	74	151	537
	%	21.4	22.5	14.2	13.8	28.1	100.0

Cramer's $V = .23$; $p = .000$.

engage in unsafe work, which showed little variation across the mines, worker response to such orders did vary significantly across the five production sites. Workers at Agrium and, to a lesser extent, Key Lake, were more likely to comply with such requests according to our data. Workers at both PCS mines were far more likely to refuse such orders and/or refer them to the shop steward or local health and safety committee. As indicated in Table 6.9, these findings are quite stark. Fewer than half as many workers at either Allan or Cory (20 to 21 per cent) showed a willingness to go along with orders that were considered risky than was the case at Agrium, where 42 per cent of those who received such work orders complied with them. As a response set, Agrium is clearly over-represented in the category of complying with supervisory requests to perform potentially dangerous work. Indeed, over a third (37 per cent) of the total consents to do jobs that were considered unsafe were registered at Agrium, and as can be seen from Table 6.9, the statistical association between worksite and type of response is a convincing one.

Are the existence of post-Fordist production relations at this site responsible for these results? Although this remains an intriguing possibility, unfortunately a definitive answer cannot be rendered. For one thing, we were unable to accurately date the incidents that were reported to us. Furthermore, and as at the other mines, unsafe work areas constitute the major category of perceived danger at Agrium. This feature seems to be more endemic to the industry than to unique managerial relations within it, although employee responses may definitely be tied up with managerial regimes. Having said that, there has been no decrease in the frequency with which serious injuries have been reported at Agrium since the movement into continuous improvement. Although the injury rate has declined at Allan since 1990, most likely propelled by the shortening of the working year for employees there, and has increased at Cory since 1987, most likely on account of the lengthening of the working year there, 1991 and 1992 were the years in which the highest number of serious accidents were reported amongst the sample at Agrium. Although hardly definitive, that more flexible work arrangements might coincide with declining propensities to refuse work on safety grounds is not a far-fetched proposal.

Related to these considerations is the issue of job stress.[35] Pressure for continuous improvement could, for example, lead to behaviours that amplify the risk-taking nature of the job. 'Management by stress' systems such as continuous improvement, which strive to incorporate stress into the organization of the labour process, are considered to be a hallmark of new, lean production methods by a number of authors.[36] Accordingly, those firms that are dedicated to new production relations might be expected to exhibit higher levels of reported stress among their labour forces. This expectation, however, is not borne out in a straightforward manner. Job stress, it seems, is a common malady that was experienced by an overall majority (57 per cent) of the total sample. Its frequency was proportionately less common at the Allan mine, where 48 per cent reported its occurrence, and most common at Central Canada, where 65 per cent of respondents reported experiencing job stress. Nevertheless, the range of positive responses to the existence of job stress is not great between the different sites. The fact that the two PCS divisions exhibit different patterns – less at Allan, more at Cory – is also instructive. This suggests that the existence of job stress owes more to local production demands, combined with an overarching commitment to obtaining 'more with less' than it does to the particular managerial strategies that stand in support of that objective. That, and the absence of any other statistically significant relationship between a particular mine and the occurrence of work-related stress does not allow us to posit a simple relationship between such new management practices as continuous improvement and more stressful work environments *in this industry*. In terms of its manifestations, the most frequently cited symptoms amongst the group of interviewees included mood swings that were portrayed as excessive irritability (18 per cent of the total sample), depression (14 per cent), chronic fatigue (11 per cent), and frequent bouts of insomnia (5 per cent).

One aspect of job safety that receives little attention is the right to work in an environment free from harassment. This is an issue that has been brought to our

TABLE 6.10
Reported job harassment

Source of harassment		Mine site					Row totals
		Allan	Central Canada	Cory	Cameco	Agrium	
By immediate supervisor							
Yes	N	32	25	12	15	14	98
	%	47.8	33.8	23.5	22.4	19.4	29.6
No	N	35	49	39	52	58	233
	%	52.2	66.2	76.5	77.6	80.6	70.4
Cramer's $V = .23$; $p = .001$.							
By senior management							
Yes	N	17	8	7	8	7	47
	%	25.8	10.8	13.7	11.9	9.7	14.2
No	N	49	66	44	59	65	283
	%	74.2	89.2	86.3	88.1	90.3	85.8
Cramer's $V = .17$; $p = .05$.							
By co-workers							
Yes	N	14	15	8	14	8	59
	%	20.9	20.3	15.7	20.9	11.1	17.8
No	N	53	59	43	53	64	272
	%	79.1	79.7	84.3	79.1	88.9	82.2
Cramer's $V = .10$; $p = .47$.							

attention by feminist scholars, often in the context of gender-stratified work-places. For decades the issue has been ignored. Now, it is in the process of being broadened to include other dimensions, including peer on peer, as well as superordinate and subordinate, and non-sexual, as well as sexual forms of demeaning harassment.[37]

We might expect that the types of work environments featured in this study, again with the possible exception of Key Lake where substantial numbers of female operatives are employed, would be relatively free of this form of conduct. The other sites are practically all-male environments and with the twin aspects of heavy equipment operation and dangerous working conditions, there is no mistaking the masculine culture that pervades the industry.[38] Nevertheless, the workers that we interviewed demonstrated a sophisticated, if often con-

TABLE 6.11
Types of reported harassment, all mines

Type	Total reported incidents (%)
Medical harassment	1.3
Intimidation by supervisor	42.6
Favouritism	6.8
Sexual	11.5
Religious	0.4
Intimidation by peer	21.7
Racism	10.2
Homophobia	5.5

tradictory, awareness of the serious issues that surround this topic. It would be delusory to pretend that sexism was not a problem, particularly among some sections of the rank and file. This is most commonly displayed in daily banter, or in some cases, possession of pornographic pictures and videos. On the other hand, 30 per cent of the respondents (mostly males) in the sample considered that they have been on the receiving end of harassing behaviour from supervisory staff. An additional 18 per cent reported on harassment from peers and 14 per cent could recall such actions coming from the ranks of more senior management. It is important to recognize that these responses reflected the interviewees own operationalized definitions of harassment – their voices reflecting upon their experiences of events that made them feel uncomfortable or left them not wanting to return to work the next day.

A glimpse at the range of these experiences was provided in a follow-up query, which asked the respondents to outline the most serious incident of harassment that either they personally had been subjected to, or had witnessed as a third party (Table 6.11). Intimidation by supervisors was the most common response, being cited by 44 per cent of those who were aware of harassment problems in their workplaces. Often this entailed verbal abuse of workers, such as swearing and other forms of demeaning address. But intimidation could also be escalated further, beyond verbal abuse. At three mine sites, workers reported on actual physical assaults by supervisory staff, such as throwing tools at workers. Less frequently mentioned were gross forms of favouritism on the part of supervisors, including patterns of inequitable work assignments, inappropriate levels of discipline, and deliberately passing over deserving workers for promotion.

Sexual harassment accounted for 12 per cent of the most serious incident recounts; racism, 10 per cent; and homophobia, 6 per cent. These practices entailed both supervisory/employee relations and worker-on-worker acts. As one might expect they are highly correlated with the composition of the work-force at each mine. Sexual harassment, up to and including allegations of sexual assault, were most frequently cited at Key Lake, which has the only significant female workforce. Even at Allan, however, where only three female workers were still employed at the time of the study, sexual harassment was mentioned by 18 per cent of the respondents, as compared with 29 per cent of the respondents at Key Lake. Racism was also most frequently raised as a concern at Key Lake, where 13 per cent of the respondents volunteered it as the most serious incident that they had been involved with, or witnessed. Overall, Key Lake accounted for one-quarter of the racist incidents that were judged to be the most serious episodes of harassment and over one-half (52 per cent) of the sexist incidents.

From the information that was collected on this most sensitive of issues, the following additional points emerge. First, with respect to direct personal experience, that is, those situations in which the interviewee was on the receiving end of the offensive behaviour, supervisory/worker episodes loom the largest. Here, there are significant differences among the mines. As found in Table 6.10, a larger proportion of the workforce at Allan (48 per cent) reported personal experience with this form of conduct. The relationship between worksite and experiences of supervisory harassment are strong and significant, with Allan, in particular, being over-represented (20 per cent of the sample with 33 per cent of the sample reporting this form of harassment). Exactly the same pattern prevails with respect to forms of harassment that are viewed as being 'directed from above,' by more senior management. Again, these are most frequently recounted at Allan and are least numerous at Agrium.

Second, although less peer-on-peer harassment was reported at Cory and Agrium, the differences with the other mines were not statistically significant. No one operation stands out in this regard. Also noteworthy is that although peer-on-peer harassment is a serious, damaging occurrence, at every mine it is less frequently experienced (although more frequently observed) than forms of harassment that entail bosses and workers.

Roughly similar patterns are registered when we consider reports of harassment that have been perpetrated on third parties and witnessed by our respondents (not shown). Peer-on-peer discrimination was witnessed much more frequently than it was actually experienced by members of the sample. With the exception of Cory, where it was reported on much less frequently, there was not much variation across the other mine sites. That Cory has a considerably smaller and more senior labour force than the other mines might account for this. Again, supervisory harassment of other workers, witnessed by respondents to this study, was most prevalent at Allan, where 67 per cent of the workers stated that they had observed this behaviour, compared with 36 per cent at Agrium. Clearly on this count, and on harassing actions that have their source at higher levels of management, PCS-Allan is over-represented, as is PCS-Cory on the latter score (that is, harassment attributed to management above the supervisory level).

That the results are consistent in a strong and statistically significant manner when it comes to supervisory or managerial harassment regardless of whether the focus is the individual respondent, or a third-party event that has been witnessed by the respondent, is indicative of a real social pattern. Management at PCS is definitely more 'old style,' and can entail intimidation, especially on the part of immediate supervisory staff. Furthermore, there is greater latitude for this type of conduct, especially at Allan, where overproduction and frequent, indefinite lay-offs are the norm. Relations between supervisory staff, higher management, and the workforce may be said to reflect the more tenuous position of the latter and the semidisposable condition in which they find themselves. These conditions, and the form of accumulation that has given rise to them, are not exactly conducive to mutual respect.

At Cory, following the radical downsizing and the adoption of niche market product specialization, the problem of overproduction was transformed into a challenge of meeting production targets and customer demand. Henceforth, the cooperation of the workforce was required for the performance of large amounts of overtime work and other forms of extra effort. Under these conditions, abusive behaviour towards workers would be a 'non-starter.' A similar dynamic might also be hypothesized as being in operation at Key Lake. The cooperation required to make radically spatialized industrial relations work may moderate the potential for abusive relationships between supervisors and workers. In addition, the race issue is a flash point at this mine and must be handled carefully. This may also have a moderating effect on first line supervision, which is not representative of the make-up of the workforce as a whole, given the complete absence of Aboriginal workers from its ranks.

At Agrium, post-Fordist work relations seemingly have less room for a 'rough' style of management. Less personal harassment was experienced and witnessed at this mine. At Central Canada, experienced levels of supervisory harassment were comparatively high, while witnessed levels were about average. That experienced levels were higher than expected might be accounted for by the halting, transitory nature of workplace reform at this mine. Management's on again, off again interest in greater employee involvement may have made a change in supervisory practices even more difficult than is normally the case. Consequently, a clear, or believable, message about change of direction may not have convinced supervisory staff of the need to change course.

The foregoing results are suggestive of two conclusions. Post-Fordist work relations do make a difference with respect to harassment in the workplace, at least as it pertains to supervisory/worker relations. The chief vector for this relationship is the need for enhanced levels of cooperation with the workforce, which is a defining feature of this mode of management. *But*, the same point can be made for any work relation that requires a greater cooperative effort; hence, the comparatively lower amounts of supervisory harassment that were also reported at Key Lake and PCS-Cory. That the levels at these sites approached the findings at Agrium suggests that the critical factor here is management's need for cooperation, as opposed to the particular package (continuous improvement or willingness to work overtime) that it comes wrapped in.

Conclusions

This chapter has sought to explore the relationships between managerial strategies and the conduct of production politics at the five mine/mill sites. Overall, post-Fordist forms of organization should not be considered a 'magic bullet' in either sense of the term. That is, they do not necessarily resolve outstanding managerial problems, as the Agrium case demonstrates so clearly. Instead of dissolving conflict into team harmony, conflict was removed from the grievance machinery of the collective agreement, resulting in disappointments that ultimately led to two strikes. Suspicion gave way to distrust.

Neither do new production relations necessarily eclipse trade unionism in the workplace. In fact, if anything in this study has been found to be damaging to union solidarity, it has been the contingent work patterns that are now practiced at PCS. And contrary to the debilitating effect which lay-offs have had at PCS, an argument could be put forward that experiments with post-Fordist arrangements at other sites have, if nothing else, stimulated debate within local unions, while sharpening responses to new managerial initiatives.

If post-Fordism has failed to resolve management's 'union problems,' or to 'put the union away,' it has also failed to deliver anything of substance to the working people that were included in this study. By no stretch of the imagination have workers been empowered by the new forms of work organization, unless by that we mean a willingness to more readily undertake dangerous work. Apart from less rough treatment at the hands of supervisory personnel, however, the differences between the mines on real empowerment issues are marginal at best. Certainly, they make it questionable as to whether old protections ought to be cashed in for new 'rights.'

7

Final Reflections

Over the course of this study, two large-scale industrial conflicts took place to which I had 'privileged' access. The first, I have already had occasion to mention, in some detail. It involved renewal of the 1994 collective agreement at Agrium. This dispute, which culminated in a one-day strike followed by agreement on a two-year extension of the existing contract (that is, status quo), was marked by a distinctive clash of vision. The union membership sought greater job security. In concrete terms, this included clear-cut language on employee seniority and its role in the workplace, and an unemployment insurance top-up program that could be called upon in the event of lay-offs. Agrium management balked at both proposals. Seniority language, which could be used to regulate lay-offs and job bidding, was rendered more difficult by the de facto payment-for-knowledge system, which was in effect at the mine. The company, for its part, was anxious that workers formally commit to a program of continuous improvement in exchange for a short-term moratorium on job loss, as well as a gainsharing program. The union remained suspicious of continuous improvement and was not assuaged by the 'guarantees' and sweeteners that the company held out as accompanying artifices.

As we saw, the gulf between these objectives was too wide, even for the joint problem-solving approach that had replaced traditional across the table negotiating. In the end, it was status quo, or more accurately, status quo ante. The former letter of understanding on continuous improvement that the union and company had signed was abrogated. The draft guidelines on continuous improvement, as well as the proposed letter of understanding on job security, were removed. The existing agreement was simply extended for another two years. The addition of some qualifying language to the application of seniority on site and a moderate wage increase constituted the only amendments.[1]

The second strike also occurred at Agrium over the spring and summer months of 1996. By then, the formal part of the study, including the component that had involved Agrium, was completed. My involvement with these events was, accordingly, less intense, and up until this point, the second dispute has played no role in the analysis. Nonetheless, one could not help but sit up and take note. New work relations, introduced to offset prior damage and to improve the existing state of affairs on site, were seeming to have just the opposite effect. Prior to the one day strike of 1994, there had only been two previous official strikes in the division's history, one in 1972 and the last in 1979.[2] Now, within two years, two strikes had been pursued by the local workforce. The second strike lasted for a startling four and a half months. This appeared to be a classic case of unanticipated consequences with a vengeance. The issues involved, and the dénouement, are instructive for what they tell us about the current state of the industry, and the post-Fordist strategies that were pursued by this one employer. These events also provide a fitting note on which to draw our discussions to a close.

On the surface, this second strike was principally about money. Agrium workers had fallen behind the rates being offered at the other mines and wanted to catch up. The union proposed a new wage schedule that was midway between the earnings of workers at Central Canada Potash and PCS.[3] With record reported company profits of $64 million and ambitious plans to significantly increase production capacity to 1.5 million tonnes in the forthcoming year, workers considered their demands to be reasonable.[4] Vouchsafing this realism, over 90 per cent of the membership voted to take strike action if their demands went unmet.[5]

Agrium countered with a monetary offer that would have left employees further behind in their industry group. The basic wage offer was, however, supplemented with a one-time-only signing bonus of $1000 and the continuation of the gainsharing program. With a $0.17 per hour differential between what the union was asking and what the company was offering, the strike was on.

Although commencing in the midst of the spring planting season, the company had a rather simple contingency plan to get it through its obligations with buyers. It would simply purchase potash from its neighbours' excess capacity to fulfill its contractual obligations and maintain its market share and customer base. The existent industrial relations of the industry allowed this to happen with not so much as a wrinkle. Thus, although Agrium was shut down for four and a half months and not one railway car of product departed from the mine's loadout area, product that now legally belonged to the company was delivered throughout the strike.[6] The results were entirely predictable.

With threats of a petition being started by a rump group within the local to force a vote on the company's last offer, the bargaining committee was prepared, 'with reservations,' to recommend the settlement after the strike had dragged on for more than four months.[7] There was no mistaking it, however; these were code words. In order to even generate a last offer from the company, the bargaining committee had had to commit to recommend it as part of the deal. This, despite the fact that the offer before the membership had not changed substantially since the company's last proposal, more than a month previously. At that time, it was rejected by a two-thirds majority.

Speeches from each individual member of the bargaining committee at the ratification meeting were also testimony to the bitter pill that was about to be swallowed. Wages in the first year of the new three-year proposal would rise to parity with the next closest producer. But after that, as renewal agreements came up in the rest of the industry, Agrium workers would once again fall to the bottom. The strike's objective of obtaining full-scale, long-standing parity with the median wage amongst the other producers had been lost. The final vote on the offer – 57 per cent in favour, 43 per cent opposed – clearly bespeaks this and the divisions that had emerged in the local. The question before us now is what lessons may be drawn from this series of events and the larger context that this study has mapped out?

In spite of earning what would generally be considered good industrial wages, there was nothing frivolous about the demands that the Agrium workers had placed before the company. At the time of our interviews, 43 per cent of the Agrium sample considered that their current wages were either just enough to get by on, or were inadequate and required another income. Almost one-quarter of the sample, 24 per cent, were in the latter category, considering a second income to be a necessity. Again, precisely one-quarter of the sample did not feel that they had adequate savings to survive economically for three months without work, even when unemployment insurance benefits were factored into their household incomes. When it became a question of subsisting for six months, this figure rose to 33 per cent of the sample who did not feel that they could make it without liquidating major assets, such as homes. Finally, a stunning 94 per cent of the sample disagreed, or disagreed strongly with the notion that their company pensions would provide an adequate share of income in their retirement years. For the workers then, there was a considerable amount at stake in this strike, much of it fuelled by the uncertainties that in part constitute post-Fordism.

Agrium workers were not alone in expressing such views, or acting upon them in individual ways. Very similar statistics could be provided for the other sites in the study. Indeed, overall, 67 per cent of those interviewed, at all of the companies, required and had second incomes in the form of working partners. The income provided by such jobs was not merely a source of 'pin' money. By and large, partners also worked full-time schedules, year-round. On top of this, practically one-third of the complete sample held down second jobs themselves, apart from their work at the various mines. What this translates into is a situation where one-fifth of all interviewees were working *three or more jobs per household* in order to make ends meet. This is one notable characteristic of the new economy: working harder for less. It is constitutive of the new treadmill of post-Fordism.

The reasons for the strike's failure are another dimension of this new economy. The transfer of production, not only from division to division within corporations, but, as in the case of this strike, from company to company, is vivid testimony to the mobility of capital. Sadly, on this occasion, competitors became willing cooperators, while co-workers remained indifferent to one another's condition. The strike merely highlighted this situation. The existing maldistribution of work within the industry and among divisions of the same company, whereby conditions of prolonged underemployment coexist with situations of overwork at other properties, provides us with another poignant reminder of the forces in the new economy that are arrayed against solidarity.

It would be arrogant to suggest that the union is not acutely aware of such problems. Small locals, with separate agreements, negotiating with major transnational corporations are clearly a mismatch. For these reasons, national representatives of the Steelworkers are pushing for the amalgamation of local unions.[8] More specifically, within potash proper, officials from the different locals of both the Steelworkers and the Communication, Energy, and Paper Workers have begun exploring the possibilities of laying the groundwork for more cooperative relations.[9] Common contract expiration dates for the industry as a whole have been suggested, so as to prevent repeats of the 1996 Agrium strike, where efforts by one local were undermined by managerial decisions taken elsewhere. This, however, would only be a start. The issue of redistributing work must also be taken on. This would entail two features, neither of which is on current agendas. First, not only must overtime work be strictly regulated, but working hours as a whole must come down. Given increases in productivity in the mining industry, a decline in working hours, without loss of income, is not an unreasonable proposal. Second, the unions themselves must be allowed input into the allocation of production between properties. This would ensure a fairer distribution of work and income than currently exists.

I am not naive. Suggestions such as these constitute real forms of empowerment, as opposed to what the post-Fordist agenda has been shown to consist of. For this reason alone, such suggestions will be ruled out-of-hand by some. However, apart from treading on management's much vaunted 'right to manage,' such proposals go directly against the current of contemporary economic trends. In today's world, even those with work often want more of it, or at least more of the income that it provides, rather than less. This is one, individual strategy that is adopted as a means of defending against the vagaries of the new economy. Workers are now more used to thinking about their place on seniority lists and the exercise of bumping rights than reduced working hours as a possible solution to current problems.[10] Local unions, for their part, even when they are stretched beyond their means, are reluctant to turn a page on local history and merge into larger, more effective units. Post-Fordism, as a regime of accumulation, plays upon and effectively uses such defensive localisms. They are created and reproduced, as it were, in the midst of the economy's own globalization.

As such, the initial reactions of working people should also be understood to be a part of the logic of the age. If this study has demonstrated one thing, it is that post-Fordism is not quite the reality that is first set out. It is both more than we may wish to bargain with, in a global sense, and less than has been advertised, in a local sense. About the empowerment of workers, households, and communities, it is not. About the creation of more participative, skilful labour processes, it is, at best, tangential. Rather, the emerging economy is, first and foremost, about doing more with less and for less. In this study, this was the one common denominator that overrode differences between individual corporate managements at the four companies, as well as occupational differences within the industry. Owing to this, the more with less phenomenon was also, in the end, the most significant factor to emerge from the investigation. Thus, despite local variations, downsizing, the expansion of work areas, and the addition of new tasks to old jobs were the real trademarks of the changes that were besetting the mining industries. These pressures were produced across the board, both by firms that visibly embraced the new discourses on involvement, improvement, and multiskilling, and by firms that chose to disregard current managerial fads. These same trends were experienced by workers in a plurality of interconnected ways; as growing employment insecurity, as work-related stress, and as a decline in quality of life. But these were also the traits that had sparked the renewal of collective struggle, which we were also given the opportunity to observe. Whether in terms of strikes, or the first tentative steps towards the re-creation of a broader solidarity in the industry, responses to the realities of the post-Fordist experience are still emerging. And in the end, like the workers at Agrium and doubtlessly countless other corporations, we remain unconvinced that such realities should proceed unchallenged.

Notes

Chapter 1: Introduction

1 Sources of information for this narrative include a newspaper clipping file that was compiled by the local union, an unofficial 'Transcript of Notes of Proceedings' made at the unfair labour practice hearings that followed the layoff and were held before the provincial Labour Relations Board (LRB), and interviews with the then current and now current local union presidents.

2 LRB File No. 159–88, Reasons for Decision in the Matter of an Application Alleging a Technological Change Pursuant to Section 43 of the Trade Union Act, R.S.S. 1978 c-T 17, 21 February 1989, 27 (hereafter, 'Reasons for Decision'); M. Marud, '200 Potash Workers to Be Laid off,' *Star-Phoenix* (Saskatoon), 25 May 1988, D-1.

3 Wages in the provincial mining industry are on a par with industrial wages in the main unionized sectors of central Canada, such as automobile manufacturing and steel production.

4 J. Burgoyne, 'Cory Workers "Can't Expect" to Be Recalled,' *Star-Phoenix*, 26 May 1988, A-1. Most potash is red in colour and comes in different grades or sizes. It is used exclusively in fertilizer production. 'White' as opposed to 'red' product contains higher quantities of potassium and is used in the manufacture of such items as pharmaceuticals, glass, soap, ammunition, and other chemicals. The decision to dedicate production at the mine to this product was given the designation of the 'white option' by company officials.

5 PCS recorded a $106 million profit in 1988 following loses of $20.7 million in 1987, $103 million in 1986, and $68 million in 1985, the three most difficult years in the company's twenty year history. *Canadian Minerals Yearbook*; Marud, '200 Potash Workers'; 'Cory Layoffs Not Appeasement to U.S., Lane Says,' *Star-Phoenix*, 30 May 1988, A-3.

6 Natural Resources Canada, *Canadian Minerals Yearbook*, 1988.

7 D. Herman, 'Miners' Wives Want Answers, Jobs Back,' *Star-Phoenix*, 31 May 1988, A-3; D. Traynor, 'Cory Layoff Plans Started Last Year,' ibid., 3 June 1988, A-1; R. Burton, 'Lanigan Mine Expansion Blamed for Cory Layoffs,' ibid., 4 June 1988, A-8.

8 United Steelworkers of America v. PCS Cory: Technological Change Application, LRB *Transcript of Notes of Proceedings* (hereafter, *Transcripts*), 68, 135.

9 J. Corman, 'The Impact of State Ownership on a State Proprietary Corporation: The Potash Corporation of Saskatchewan,' (PhD dissertation, University of Toronto, 1982).

10 In addition to the *Transcripts* this narrative is also based upon the 'Legal Brief' filed by the United Steelworkers Local 7458, 18 October 1988, 'Written Argument of the Respondent,' (PCS), 26 October 1988; and 'Reasons for Decision,' LRB File No. 159–88, 21 February 1989.

11 *Revised Statutes of Saskatchewan*, 1978, Chapter T-17, 6179.

12 *Transcripts*, 18, 67–8, 129–33, 154–5.

13 Ibid., 9, 139–44, 148, 150, passim.

14 Ibid., 10.

15 Written Argument Filed on Behalf of the Respondent, LRB File 159–88, 19.

16 USWA LU 7689 past President, interview, 22 April 1995.

17 USWA Staff Representative District 3, interview, 5 June 1995.

18 Ibid. The failure of either the USWA or the CEP to gain the certification at PCS-Rocanville, which is now represented by an independent employee association, has been attributed to such competition.

19 The 1987 Lanigan strike was a particularly bitter affair, with mass picketing, an injunction, and a large-scale police presence that included over sixty arrests in one incident. Significantly, original union demands focused on a guarantee of no job losses due to technological change, a 5 per cent wage increase over two years, and continuation of existing holiday premiums for those on twelve-hour shifts. In each respect, the workers were unsuccessful, settling for a wage offer that had previously been rejected (6 per cent over four years plus signing bonus), and no company movement on the other issues. (D. Herman, 'Potash Union Sued over Blockade at Lanigan Mine,' *Star-Phoenix*, 13 March 1986, A-8; D. Herman, 'Lanigan Workers Welcome Pact,' ibid., 30 January 1987, A-1; D. Herman, 'Potash Pact Brings Peace to Lanigan,' ibid., 4 February 1987, A-3.

20 The company took the position that a severance package was in the works, but that the union's unfair labour practice suit removed the good will that would have been required to implement it. This strikes one as disingenuous. If the company had a severance package in preparation why would it not have unveiled it at the time of the lay-off announcement, or shortly thereafter, and avoided the negative publicity associated with the lay-off?

21 D. Drache and H. Glasbeek, *The Changing Workplace: Reshaping Canada's Industrial Relations System* (Toronto: James Lorimer, 1992); L. Haiven, 'Hegemony and the Workplace: The Role of Arbitration,' in *Regulating Labour*, eds. S. McBride et al. (Toronto: Garamond, 1991).

22 Local Union President, interview, 15 October 1995; District Staff Representative, interview, 18 October 1995.

23 'Reasons for Decision,' LRB File No. 159–88, 21 February 1989.

24 Saskterra, meanwhile, was a subsidiary of Husky Oil Ltd. With this acquisition, PCS became the sole owner of the Allan facility.

25 Phosphate is one of the main nutrients in commercial fertilizer along with potash and nitrogen. According to the business press, 'the deal gives Potash Corp. the ability to exploit its marketing expertise by selling phosphate to many of the same customers already buying its potash.' Robinson, 'Potash Corp. to Acquire Texasgulf,' *Globe and Mail*, 7 March 1995, B-1.

26 P. Fritsch, 'Potash Corp. to Buy Arcadian,' *Globe and Mail*, 3 September 1996, B-1.

27 A. Freeman, 'Potash Corp. Seeks Global Markets,' *Globe and Mail*, 29 August 1996, B-1.

28 Ibid.

29 On the interconnectedness of some of these dimensions see D. Harvey, *The Condition of Postmodernity* (Oxford: Basil Blackwell, 1989); G. Arrighi, *The Long Twentieth Century* (London: Verso, 1995).

30 A clear, if controversial statement of these trends is contained in R. Reich, *The Work of Nations* (New York: Vintage Books, 1991).

31 A. Lipietz, *Mirages and Miracles* (London: Verso, 1987) begins to undertake this type of analysis.

32 D. Bell, *The Coming of Post-Industrial Society* (New York: Basic Books, 1973); F. Block, *Post-Industrial Possibilities* (Berkeley: University of California Press, 1990).

33 S. Cohen and J. Zysman, *Manufacturing Matters: The Myth of the Post-Industrial Economy* (New York: Basic Books, 1987).

34 For the emergence of a new assembly line Fordist industry/service, see E. Reiter, *Making Fast Food* (Montreal: McGill-Queen's University Press, 1991).

35 H. Braverman, *Labor and Monopoly Capital* (New York: Monthly Review Press, 1974).

36 See for example L. Hirschhorn, *Beyond Mechanization: Work and Technology in a Postindustrial Age.* (Cambridge Mass: MIT Press, 1984). A contemporary sociotechnical application can be found in T. Rankin, *New Forms of Work Organization.* (Toronto: University of Toronto Press, 1990). These comments are expanded upon in Chapter 5.

37 A. Gramsci, *Selections from The Prison Notebooks* (London: Lawrence and

Wishart, 1971); M. Aglietta, *A Theory of Capitalist Regulation* (London: New Left Books, 1979); R. Boyer, *The Regulation School: A Critical Introduction* (New York: Columbia University Press, 1990); A. Lipietz, *The Enchanted World* (London: Verso, 1985); and Lipietz, *Mirages and Miracles.*

38 J. Jenson, '"Different" but Not "Exceptional": Canada's Permeable Fordism,' *Canadian Review of Sociology and Anthropology* 26, no. 1 (1989); B. Jessop, *State Theory* (Oxford: Polity Press, 1990); B. Russell, *Back to Work? Labour, State, and Industrial Relations in Canada* (Scarborough: Nelson, 1990).

39 M. Dubofsky, *The State and Labor in Modern America* (Chapel Hill: University of North Carolina Press, 1994); K. Klare, 'Judicial Deradicalization of the Wagner Act and the Origins of Modern Legal Consciousness,' *Minnesota Law Review* 62 (1978); C. Tomlins, *The State and the Unions* (Cambridge: Cambridge University Press, 1985); G. Baglioni and C. Crouch, eds., *European Industrial Relations: The Challenge of Flexibility* (London: Sage, 1990); A. Ferner and R. Hyman, eds., *Industrial Relations in the New Europe* (Oxford: Basil Blackwell, 1992).

40 R. Edwards, *Contested Terrain* (New York: Basic Books, 1979); S. Jacoby, *Employing Bureaucracy* (New York: Columbia University Press, 1985); H. Katz, *Shifting Gears* (Cambridge: MIT Press, 1985).

41 D. Brody, *Workers in Industrial America* (New York: Oxford University Press, 1980); H. Harris, *The Right to Manage* (Madison: University of Wisconsin Press, 1982); K. Klare, 'Judicial Deradicalization'; N. Lichtenstein, *Labor's War at Home* (Cambridge: Cambridge University Press, 1982); D. Montgomery, *Worker's Control in America* (Cambridge: Cambridge University Press, 1980); J. Stepan-Norris and M. Zeitlin, '"Red" Unions and "Bourgeois" Contracts,' *American Journal of Sociology* 96, no. 5 (1991).

42 Lipietz, *Mirages and Miracles.*

43 Braverman, *Labor and Monopoly Capital.*

44 R. Cox, *Production, Power, and World Order: Social Forces in the Making of History* (New York: Columbia University Press, 1987).

45 A. Johnson, S. McBride, and P. Smith, eds., *Continuities and Discontinuities: The Political Economy of Social Welfare and Labor Market Policy in Canada* (Toronto: University of Toronto Press, 1994).

46 G. Therborn, *Why Some People Are More Unemployed Than Others* (London: Verso, 1986).

47 Russell, *Back to Work*, 182.

48 G. Esping-Andersen, *The Three Worlds of Welfare Capitalism* (Princeton: Princeton University Press, 1990); G. Olsen, 'Locating the Canadian Welfare State,' *Canadian Journal of Sociology* 19, no. 1 (1994).

49 N. Poulantzas, *State, Power, Socialism* (London: Verso, 1980).

50 Boyer, *The Regulation School.*

51 Arrighi, *The Long Twentieth Century.*

52 R. Parboni, *The Dollar and Its Rivals: Recession, Inflation, and International Finance* (London: New Left Books, 1981); F. Block, *The Origins of International Economic Disorder* (Berkeley: University of California Press, 1977).

53 A. Gamble, *The Free Economy and the Strong State: The Politics of Thatcherism* (London: Macmillan, 1988); B. Jessop et al., *Thatcherism* (Oxford: Polity Press, 1988); B. Harrison and B. Bluestone, *The Great U-Turn: Corporate Restructuring and the Polarizing of America* (New York: Basic Books, 1988).

54 S. McBride, *Not Working: State, Unemployment and Neo-Conservatism in Canada* (Toronto: University of Toronto Press, 1992).

55 S. McBride and J. Shields, *Dismantling a Nation* (Halifax: Fernwood, 1993).

56 The main components of such debt are not welfare charges, but interest payments owing on the debt. These exorbitant charges remain the legacy of the monetarist experiment.

57 M. Piore and C. Sabel, *The Second Industrial Divide* (New York: Basic Books, 1984).

58 Braverman, *Labor and Monopoly Capital*; also, D. Nelson, *Fredrick W. Taylor and the Rise of Scientific Management* (Madison: University of Wisconsin Press, 1980); D. Clawson, *Bureaucracy and the Labor Process* (New York: Monthly Review Press, 1980).

59 D. Noble, *Forces of Production* (New York: Oxford University Press, 1986).

60 M. Burawoy, *Politics of Production* (London: Verso, 1985); R. Edwards, *Contested Terrain*; P. Edwards, *Conflict at Work* (Oxford: Basil Blackwell, 1986); A. Friedman, *Industry and Labour* (London: Macmillan, 1977); D. Knights and H. Willmott, *Labour Process Theory* (London: Macmillan, 1990); C. Littler, *The Development of the Labour Process in Capitalist Societies* (London: Heinemann, 1982); D. Nelson, 'Scientific Management and the Workplace,' in *Masters to Managers*, ed. S. Jacoby (New York: Columbia University Press, 1991); B. Russell, 'The Subtle Labour Process and the Great Skill Debate,' *Canadian Journal of Sociology*, 20, no. 3 (1995).

61 J. Fudge, 'Voluntarism and Compulsion: The Canadian Federal Government's Intervention in Collective Bargaining from 1900 to 1946' (PhD dissertation, University College, Oxford University 1987); D. Millar, 'The Shapes of Power: The Ontario Labour Relations Board, 1944 to 1950' (PhD dissertation, York University, 1980); P. Warrian, 'Labour Is Not a Commodity' (PhD dissertation, University of Waterloo, 1986).

62 J.M. Keynes, *Essays in Persuasion* (London: Macmillan, 1953). In Canada, the most important Keynesian charter documents consist of the *Report of the Royal Commission on Dominion–Provincial Relations*, Book 1 and 2 (Ottawa: King's Printer, 1940) and the Minister of Reconstruction, *Employment and Income* (Ottawa: King's Printer, 1945).

63 Harvey, *The Condition of Postmodernity*; Piore and Sabel, *Second Industrial Divide*.

64 M. Kenney and R. Florida, 'Beyond Mass Production: Production and the Labor Process in Japan,' *Politics and Society* 16, no. 1 (1988); Kenney and Florida, *Beyond Mass Production* (New York: Oxford University Press, 1993).

65 R. Jaikumar, 'Postindustrial Manufacturing,' *Harvard Business Review* 6 (1986); Piore and Sabel, *Second Industrial Divide*; J. Womack et al., *The Machine That Changed the World* (New York: Rawson Associates, 1990).

66 Braverman, *Labor and Monopoly Capital*; Edwards, *Contested Terrain*; D. Noble, *Progress Without People* (Toronto: Between the Lines Press, 1995); M. Burawoy, *Manufacturing Consent* (Chicago: University of Chicago Press, 1979); Buroway, *Politics of Production*; Jacoby, *Employing Bureaucracy*.

67 Edwards, *Conflict at Work*; Russell, 'Subtle Labour Process.'

68 J. Atkinson, 'Manpower Strategies for Flexible Organisations,' *Personnel Management* (1984).

69 Katz, *Shifting Gears*.

70 Kenney and Florida, *Beyond Mass Production*.

71 T. Kochan et al., *The Transformation of American Industrial Relations* (New York: Basic Books 1986); Katz, *Shifting Gears*.

72 C. Heckscher, *The New Unionism* (New York: Basic Books, 1988).

73 McBride and Shields, *Dismantling a Nation*; T. Dunk, S. McBride, and R. Nelson, eds., *The Training Trap: Ideology, Training, and the Labour Market* (Halifax: Fernwood, 1996).

74 R. Mahon, 'From Fordism to ? New Technology, Labour Markets, and Unions,' *Economic and Industrial Democracy* 8, no. 1, 1987; Boyer, *The Regulation School*. Also see W. Lazonick, *Business Organization and the Myth of the Market Economy* (Cambridge: Cambridge University Press, 1991).

75 D. Drache, 'The Way Ahead for Ontario' and D. Wolfe, 'Technology and Trade' in *Getting on Track: Social Democratic Strategies for Ontario*, ed. D. Drache (Montreal: McGill-Queen's University Press, 1992).

76 Piore and Sabel, *Second Industrial Divide*; M. Piore, 'Perspectives on Labor Market Flexibility,' *Industrial Relations* 25, no. 2 (1986).

77 Specifically, Piore and Sabel allow that 'flexible specialization opens up long-term prospects for improvement in the conditions of working life – regardless of this system's effect on the balance of power between currently existing organizations of capital and labour.' For this reason there 'is a case for preferring it [flexible specialization] to mass production, regardless of the place accorded to unions within craft production.' Piore and Sabel, *Second Industrial Divide*, 278; also see Katz, *Shifting Gears*, for a similar position.

78 D. Coates, *The Crisis of Labour* (Oxford: Philip Allan, 1989); J. Kelly, *Trade*

Unions and Socialist Politics (London: Verso, 1988); J. MacInnes, *Thatcherism at Work* (Milton Keynes: Open University Press, 1987); A. Pollert, 'The Flexible Firm,' *Work, Employment and Society* 2, no. 3 (1988); and 'Dismantling Flexibility,' *Capital and Class* 34 (1988).

79 J. Price, 'Lean Production at Suzuki and Toyota,' *Studies in Political Economy* 45 (1994); J. Rinehart, 'Improving the Quality of Work through Job Redesign,' *Canadian Review of Sociology and Anthropology* 23, no. 4 (1986); H. Shaiken et al., 'The Work Process under More Flexible Production,' *Industrial Relations* 25, no. 2 (1986); D. Swartz, 'New Forms of Worker Participation,' *Studies in Political Economy* 5 (1981).

80 C. Berggren, *Alternatives to Lean Production* (Ithaca, NY: ILR Press, 1992); J. Corman et al., *Recasting Steel Labour* (Halifax: Fernwood, 1993); K. Dohse et al., 'From "Fordism" to "Toyotism": The Social Organization of the Labor Process in the Japanese Automobile Industry,' *Politics and Society* 14, no. 2 (1985); D. Drache and H. Glasbeek, *The Changing Workplace*; S. Wood, 'The Transformation of Work?' in *The Transformation of Work?* ed. S. Wood (London: Unwin Hyman, 1989).

81 Kenney and Florida, *Beyond Mass Production*.

82 See the works by L. Panitch and D. Swartz: 'Towards Permanent Exceptionalism,' *Labour/Le Travail* (1984); *From Consent to Coercion* (Toronto: Garamond, 1985); *The Assault on Trade Union Freedoms* (Toronto: Garamond, 1988); *The Assault on Trade Union Freedoms: From Wage Controls to Social Contract* (Toronto: Garamond, 1993); and H. Glasbeek, 'Labour Relations Policy and Law as Mechanisms of Adjustment,' *Osgoode Hall Law Journal* 25, no. 1 (1987).

83 Panitch and Swartz, *Assault on Trade Union Freedoms*; S. McBride, 'Hard Times and "The Rules of the Game,"' in *Working People and Hard Times*, ed. R. Argue et al. (Toronto: Garamond, 1987).

84 D. Drache and H. Glasbeek, 'The New Fordism in Canada: Capital's Offensive, Labour's Opportunity,' *Osgoode Hall Law Journal* 27, no. 3 (1989); *The Changing Workplace*; B. Russell, 'Labour's Magna Carta? Wagerism in Canada at Fifty,' in *Labour Pains, Labour Gains: 50 Years of PC1003*, ed. Gonick et al. (Halifax: Fernwood, 1995); T. Wannell, 'Firm Size and Employment: Recent Canadian Trends,' *Canadian Economic Observer* (1992).

85 On this point, see the collection of papers in Gonick et al., *Labour Pains, Labour Gains*.

86 J. Jenson and R. Mahon, 'North American Labour: Divergent Trajectories' and 'Legacies for Canadian Labour of Two Decades of Crisis,' in *The Challenge of Restructuring: North American Labor Movements Respond*, ed. J. Jenson and R. Mahon (Philadelphia: Temple University Press, 1993); R. Freedman and J. Medoff, *What Do Unions Do?* (New York: Basic Books, 1984).

87 Price, 'Lean Production.'

88 D. Wells, *Empty Promises: Quality of Work Life Programs and the Labor Movement* (New York: Monthly Review Press, 1987).

89 Rankin, *New Forms of Work Organization*; E. Trist et al., 'An Experiment in Autonomous Working in an American Underground Coal Mine,' *Human Relations* 30, no. 3 (1977).

90 Drache and Glasbeek, 'The New Fordism'; Jenson, '"Different" but Not "Exceptional"'; R. Mahon, 'Post-Fordism: Some Issues for Labour,' in *The New Era of Global Competition: State Policy and Market Power*, ed. D. Drache and M. Gertler, (Montreal: McGill-Queen's University Press, 1991); Jenson and Mahon, 'Legacies for Canadian Labour.'

91 W. Clement, 'Debates and Directions: A Political Economy of Resources,' in *The New Canadian Political Economy*, ed. W. Clement and G. Williams (Montreal: McGill-Queen's University Press, 1989).

92 M. Luxton, *More Than a Labour of Love* (Toronto: Women's Press 1980); N. Denis et al., *Coal Is Our Life* (London: Eyre and Spottiswoode, 1956).

93 Previous sociological studies of the mining industry of note include, W. Clement, *Hardrock Mining* (Toronto: McClelland and Stewart, 1981); Denis et al., *Coal Is Our Life*; A. Gouldner, *Patterns of Industrial Bureaucracy* (New York: Free Press, 1964); *Wildcat Strike* (Yellow Springs, OH: Antioch Press, 1964); E. Trist et al., *Organizational Choice* (London: Tavistock, 1963).

94 For a representative literature see, C. van Onselen, *Chibaro: African Mine Labour in Southern Rhodesia, 1900–1933* (London: Pluto Press, 1976); H. Wolpe, 'Capitalism and Cheap Labour-Power in South Africa,' *Economy and Society* 1, no. 4 (1972); M. Burawoy, 'The Functions and Reproduction of Migrant Labor: Comparative Material from Southern Africa and the United States,' *American Journal of Sociology* 81, no. 5 (1976); M. Murray, ed., *South African Capitalism and Black Political Opposition*, part 3 (Cambridge, MA: Schenkman Publishing, 1982).

95 Harvey, *The Condition of Postmodernity*.

96 In 1988 Eldorado Nuclear was merged with the Saskatchewan Mining Development Corporation to form Cameco. In 1991 the province sold off a majority of its shares on the private market, thereby taking it out of state hands.

97 P. Edwards and H. Scullion, *The Social Organization of Industrial Conflict* (Oxford: Basil Blackwell, 1982); P. Edwards, *Managing the Factory* (Oxford: Basil Blackwell, 1987); L. Haiven, 'The Political Apparatuses of Production: Generation and Resolution of Industrial Conflict in Canada and Britain' (PhD dissertation, University of Warwick, 1988).

98 Cominco potash had until 1993 been a division of the much larger Cominco Metals Company, a large Canadian multinational that specialized in hard rock mining.

In that year the potash division became a stand alone entity with its own share capital base. In 1995 the name was changed from Cominco Fertilizers Ltd. to Agrium Inc.

99 United Steel Workers of America, 'Steelworker Guidelines for Participation in Work Reorganization' (conference, 1992, mimeographed).

100 A sixth mine represented by the Steelworkers does not employ underground miners, but instead uses a forced high pressure brine extraction method, commonly referred to as solution mining. As this method does not employ many workers, and as this particular mine does not account for a very large share of production, it was not included in this study.

101 All four potash sites are underground mines. The Key Lake uranium mine is an open-pit operation. On the problem of the work-effort bargain see the original formulation by W. Baldamus, *Efficiency and Effort* (London: Tavistock, 1961).

102 Shortly after the commencement of the study, the acting union staff representative was replaced by a new permanent representative. Although this individual 'inherited' the project, I continued to receive his utmost cooperation and support.

103 For another valuable discussion of workplace research strategies that employs a somewhat different tack see R. Milkman, *Farewell to the Factory* (Berkeley: University of California Press, 1997), Appendix 3.

Chapter 2: Market Preliminaries

1 K. Polanyi, *The Great Transformation* (Boston: Beacon Press, 1957).

2 D. Drache and W. Clement, *The New Practical Guide to Canadian Political Economy* (Toronto: James Lorimer, 1985); M. Watkins, 'The Political Economy of Growth,' and W. Clement, 'Debates and Direction: A Political Economy of Resources,' both in *The New Canadian Political Economy.*

3 H. Innis, *The Fur Trade* (Toronto: University of Toronto Press, 1977); V. Fowke, *The National Policy and the Wheat Economy* (Toronto: University of Toronto Press, 1978); M. Watkins, 'A Staple Theory of Economic Growth,' in *Approaches to Canadian Economic History*, ed. W.T. Easterbrook and M.H. Watkins (Toronto: McClelland and Stewart, 1967).

4 W. Clement, *Continental Corporate Power* (Toronto: McClelland and Stewart, 1977); T. Naylor, *The History of Canadian Business, 1867–1914*, 2 vols. (Toronto: Lorimer, 1975); W. Carroll, 'The Canadian Corporate Elite: Financiers or Finance Capitalists,' *Studies in Political Economy* 8 (1982); *Corporate Power and Canadian Capitalism* (Vancouver: UBC Press, 1986); J. Niosi, *Canadian Capitalism: A Study of Power in the Canadian Business Establishment* (Toronto: Lorimer, 1981); L. Panitch, 'Dependency and Class in Canadian Political Economy,'

Studies in Political Economy 6 (1981); P. Resnick, 'From Semiperiphery to Perimeter of the Core: Canada's Place in the Capitalist World Economy,' *Review* 12, no. 2 (1989).

5 This status was also held prior to 1970, when Soviet production for the first time surpassed Canadian. J. Richards and L. Pratt, *Prairie Capitalism: Power and Influence in the New West* (Toronto: McClelland and Stewart, 1979), 194. Currently it is estimated that only 43 per cent of the FSU production capacity is being utilized, while recently the two Russian potash producers have been privatized. *Canadian Minerals Yearbook* (Ottawa: Natural Resources Canada, 1994).

6 Richards and Pratt, *Prairie Capitalism*, 138.

7 For details on the new taxes that the NDP government levied on the potash producers and the concerted industry boycott of these measures see M. Molot and J. Laux, 'The Politics of Nationalization,' *Canadian Journal of Political Science* 12, no. 2 (1979); also Richards and Pratt, *Prairie Capitalism*, ch. 10.

8 One tonne (metric) is equal to 1.1 ton (short), or 2,205 pounds.

9 Richards and Pratt's work on the new state entrepreneurship represents an interesting case. Although the authors correctly signal the importance of unstable commodity prices in provincial politics (for example, potash in 1960s Saskatchewan), in their policy prescriptions they seem to ignore this record in advocating a staples-led economic strategy under state tutelage. Rejecting the implications of an Innisian analysis, they nonetheless employ many of the staple school's analytical assumptions, including the importance of an 'entrepreneurial culture.' On the prairies, this is apparently to be provided by the state, either in right populist (Alberta) or social democratic (Saskatchewan) guise. Richards and Pratt, *Prairie Capitalism*.

10 For details see J. Pitsula and K. Rasmussen, *Privatizing a Province: The New Right in Saskatchewan* (Vancouver: New Star Books, 1990), ch. 10.

11 The conversion into nuclear fuel is carried out at refineries in Blind River and Port Hope, Ontario. At Blind River, yellow cake is refined into uranium trioxide (UO_3), an intermediate product. At Port Hope, the UO_3 is converted into uranium hexafluoride (UF_6) for use in light water reactors, or it is processed into uranium dioxide (UO_2), as fuel for the CANDU reactor.

12 . Eldorado Mining and Refining Company was created as a federal crown corporation in 1944, following the acquisition of a private company of the same name. The Port Hope refinery was also acquired in this deal. C. Giangrande, *The Nuclear North* (Toronto: Anansi, 1983); *Saskatchewan Government Resource Series, Uranium* (Regina: Saskatchewan Education, 1990).

13 B. Harding, 'The Two Faces of Public Ownership: From the Regina Manifesto to Uranium Mining,' in *Social Policy and Social Justice: The NDP Government in Saskatchewan during the Blakeney Years*, ed. J. Harding (Waterloo: Wilfrid Laurier University Press, 1995).

14 Ibid., 295.

15 Giangrande, *The Nuclear North*, 88.

16 D. Gullickson, 'Uranium Mining, the State, and Public Policy in Saskatchewan, 1971–1982: The Limits of the Social Democratic Imagination' (MA thesis, University of Regina, 1990), 68.

17 Ibid., 73, 75.

18 The bid to earn hard currency led to republics of the FSU being charged with the dumping of uranium on the U.S. market in 1991. This has been followed by the signing of export restraint agreements (quotas). Adding to the supply of uranium has also been the decommissioning of the FSU's nuclear arsenal, whereby highly enriched uranium from nuclear warheads is to be blended down into lowly enriched uranium for commercial use.

19 In the early 1990s 61 per cent of Ontario Hydro's output was generated through nuclear stations. No other province comes close to matching this level of reliance on the nuclear option.

20 *Canadian Minerals Yearbook*, 1994.

21 Saskatchewan Bureau of Statistics, 'Saskatchewan Economic Statistics' (Regina: 1994, mimeographed).

22 *Manufacturing Industries of Canada: National and Provincial Areas* (Ottawa: Statistics Canada, 1993), cat. 31–203.

23 Ibid.

24 These are company totals. They include mine and mill operators, skilled maintenance workers, technical, professional, and office staff. *Canadian Minerals Yearbook*, 1994.

25 At some mines, clerical and technical staff (e.g., lab analysts) are part of the bargaining unit, while at other sites they remain out of scope. For purposes of comparability this study includes only miners, mill employees, maintenance workers, and employees who are assigned load out, warehousing, and similar tasks. The employment figure given in the text reflects this decision, as does the data to be presented on uranium mining.

26 M. Luxton and J. Corman, 'Getting to Work: The Challenge of the Women Back into Stelco Campaign,' *Labour/Le Travail* 28 (1991).

27 Marchack also discusses this problem. P. Marchack, *Green Gold: The Forest Industry in British Columbia* (Vancouver: UBC Press, 1983).

28 The current collective agreement between Cameco and USWA 8914 contains one article with six clauses on the issue of affirmative action/employment equity. Clause 1 states: 'The union acknowledges the Company has entered into a Surface Lease Agreement with the Province of Saskatchewan which has as one of its objectives, the maximizing of employment of residents of Saskatchewan's north. The Union further acknowledges the Company has as one of its objectives, the maxi-

mizing of employment of northern residents of aboriginal ancestry.' To this Clause 5 adds: 'Notwithstanding ... it is understood and agreed that no employee who was hired prior to November 16, 1993 shall have his seniority rights affected in matters relating to a reduction in the work force or recall, as a result of preference being given to northern residents of aboriginal ancestry.' Collective Agreement between Cameco Corporation and USWA Local 8914, 1993.

29 Cameco Corporation, *1994 Annual Report*, 25.

30 Our total workforce sample for this study was 331, which represents 29.3 per cent of the relevant population at the five mine sites. This included a sample of 67 workers at Key Lake and 264 employees at the four potash sites. Also see note no. 25 for the occupational groups that are included in the sample.

31 Cameco Corp., *1994 Annual Report*.

32 A company official estimates that 35 to 37 per cent of the company's northern workers have migrated and now live in the south. Cameco Manager of Northern Affairs, Training and Education, interview, 19 January 1996.

33 Cameco Corp., *1994 Annual Report*, 19. Aboriginal employees are defined as 'persons who are Indians, Inuit, or Metis and who at the time of hire, identify themselves as such to the Company.' Collective Agreement Between Cameco Corporation and United Steel Workers of America Local 8914, 1993.

34 Uranium City was abandoned in the 1980s. There is no mining currently being undertaken around this former settlement.

35 Northern Educational Services Branch, Training for the Mineral Sector: Overview of the Multi-Party Training Plan, n.d.; Cameco Mine Manager, interview, 10 January 1996; Cameco Manager of Northern Affairs, Training and Education, interview, 19 January 1996.

36 Cameco Manager of Northern Affairs, Training and Education, interview, 19 January 1996. Since start up, this manager estimates that as many as 1500 workers of aboriginal descent have worked for the company.

Chapter 3: Corporate Cultures of Employment I

1 K. Marx, *Capital*, vol. 3 (Moscow: Progress Publishers, 1971), 791.

2 Ibid., 792.

3 Gramsci, *Selections from the Prison Notebooks*, 302.

4 Ibid., 285.

5 Ibid., 286, 296–7, 300–4.

6 J. Corman, 'The Impact of State Ownership on a State Proprietary Corporation.'

7 Compensation paid to the former private owners for their assets has been described as 'extremely generous' by Richards and Pratt, *Prairie Capitalism*, 271.

8 Potash Corporation of Saskatchewan Inc., *1991 Annual Report*, 5.

9 P. Fritsch, 'Potash Corp. to Buy Arcadian,' *Globe and Mail*, 3 September 1996, B-1; G. Ip, 'Potash Corp. Building Empire,' ibid., 4 September 1996, B-1.

10 Potash Corporation of Saskatchewan Inc., *1994 Annual Report.*

11 Ibid. and Corman, *The Impact of State Ownership*, 215.

12 A. Freeman, 'Potash Corp. Seeks Global Markets,' *Globe and Mail*, 29 August 1996, B-1; A. Freeman, 'Potash Gets Deal for Kali and Salz,' ibid., 10 December 1996, B-7; B. Milner, 'Potash Corp. Ends Global Buying Spree,' ibid., 12 December 1996, B-2.

13 The new work environment board was provided with a mandate to 'consider possible improvements to the psycho-social and physical aspects of the work environment.' Corman, *The Impact of State Ownership*, 311. Also see D. Bobiash, 'Attempts to Promote Industrial Democracy in the Potash Corporation of Saskatchewan' (MSc research paper in Industrial Relations and Personnel Management, London School of Economics, 1984), who details the controversies between management and labour on the newly created board.

14 Corman suggests that the unions in the potash industry did not show much interest in industrial democracy but she does not really distinguish between the Quality of Working Life programs that were in vogue at the time and more genuine forms of empowerment. Sass, on the other hand, provides some clarification arguing that the unions were interested in a genuine democratization of the workplace, but were definitely opposed to the QWL style reforms that were acceptable to the management of the crown corporation. See R. Sass, 'The Work Environment Board and the Limits of Social Democracy,' in *Social Policy and Social Justice*, ed. Harding.

15 See J.K. Laux and M.A. Molot, *State Capitalism: Public Enterprise in Canada* (Ithaca: Cornell University Press, 1988), 109–13.

16 Corman, 'The Impact of State Ownership,' 252; also see Pitsula and Rasmussen, *Privatizing a Province*, 94.

17 Laux and Molot, *State Capitalism*, 113.

18 Also see B. Russell, 'The Subtle Labour Process and the Great Skill Debate: Evidence from a Potash Mine-Mill Operation,' for a detailed analysis of managerial forms of labour control at the Allan site.

19 On the use of this type of training in the mining industry also see W. Clement, *Hardrock Mining*, 288. As Clement succinctly notes, 'The training unit is the equipment, not the person.'

20 Potash Corporation of Saskatchewan, NM-05:1993.

21 Ibid.

22 Potash Corporation of Saskatchewan Inc., *1994 Annual Report*, 14.

23 However, Hall notes that Loss Control was originally designed for the South African mining industry. A. Hall, 'Production Politics and the Construction of

Consent: The Case of Health and Safety in Mining' (PhD dissertation, University of Toronto, 1989), 477.

24 International Loss Control Institute, *Mine Safety and Loss Control* (Loganville, GA: Institute Press, 1984), 1.

25 F. Bird, *Mine Safety and Loss Control Management* (Loganville, GA: Institute Press, 1980), 20.

26 F. Bird and R. Loftus, *Loss Control Management* (Atlanta: Institute Press, 1976), 57.

27 Bird, *Mine Safety*, 68.

28 Ibid., 82. Hall also observes the 'increase in direct monitoring and reporting of worker's activities by a growing "safety" bureaucracy' at Inco in Sudbury after it adopted a Loss Control program at about the same time as PCS did. Hall, 'Production Politics and the Construction of Consent,' 479.

29 Hall, 'Production Politics and the Construction of Consent,' 480; also A. Hall, 'The Corporate Construction of Occupational Health and Safety: A Labour Process Analysis,' *Canadian Journal of Sociology* 18, no. 1 (1993). Hall provides the only other sociological analysis of Loss Control that I am aware of. Although I agree with his analysis on most points, he seems to overlook the strong affinities that Loss Control has with scientific management.

30 Following the Allan strike, which was fought over the introduction of CWS, the system was brought in at other company divisions, including Cory without further industrial action. For historical accounts of CWS see, R. Bean, 'The "Cooperative Wage Study" and the Canadian Steelworkers,' *Relations Industrielles* 19, no. 1 (1964); S. Jacoby, *Employing Bureaucracy*; and R. Storey, 'The Struggle for Job Ownership in the Canadian Steel Industry,' *Labour/Le Travail* 34 (1994).

31 *Co-operative Wage Study Manual* (n.d., mimeographed), 20.

32 Ibid., 6.

33 On this concept and its importance in North American industrial relations, see H. Katz, *Shifting Gears*, 39–40 and T. Kochan et al., *The Transformation of American Industrial Relations*, 28–9. Edwards' notion of bureaucratic control also runs parallel with and overlaps the concept of job control unionism, R. Edwards, *Contested Terrain*, ch. 8.

34 CWS trainee, interview, 7 July 1993.

35 Over the succeeding years, the job class differential has been expanded to its present $0.44. PCS-Allan, 1983–85 Agreement; 1995–97 Agreement.

36 CWS trainee, interview, 7 July 1993.

37 LU7689 President, interview, 14 July 1993.

38 The company's point person in charge of CWS was subsequently dismissed in the aftermath of the strike. According to one local union official, the company went

into CWS blind. Only after agreeing to it in principle and costing it out following the job evaluations did management realize the implications. LU7689 President, interviews, 14 July 1993, and 4 February 1996.

39 Both Clement and Hall inexplicably equate CWS with an enhancement of job responsibilities. See W. Clement, *Hardrock Mining: Industrial Relations at Inco*, 292–3; and Hall, 'Production Politics and the Construction of Consent: The Case of Health and Safety in Mining,' 381.

40 PCS-Allan General Mine Manager, interview, 4 November 1993; Agrium General Mine Manager, interview, 1 November 1994; and Central Canada Potash General Mine Manager, interview, 3 May 1995. Also see note 36.

41 One tonne (metric) equals 1.1 ton (short) or 2,205 pounds.

42 Potash Corporation of Saskatchewan. *1994 Annual Report*, 11.

43 Ibid., 1.

44 Ibid., 21.

45 According to one source, PCS sells its product for $0.10 more per tonne than the competition. Although this means that a small part of the market is given over to the competition, company profits are maximized. LU7689 past President, interview, 7 February 1996.

46 Pitsula and Rasmussen, *Privatizing a Province*, 93.

47 LU 7689 past President, interview, 6 January 1994.

48 At other mine sites where the lay-off language differed from Allan, lay-offs were shorter, but more frequent. LU7689 President, interview, 21 May 1993.

49 PCS-Allan respondent, interview, 29 June 1993.

50 For a discussion of the concept of subemployment see D. Livingstone, 'Post-Industrial Dynamics: Economic Restructuring, Underemployment and Lifelong Learning in the Information Age' (Twenty-seventh Annual Sorokin Lecture, University of Saskatchewan, 1996, mimeograph). Among the costs borne by workers, including some of the interviewees, are lost homes and a generally lower standard of living.

51 PCS-Allan General Mine Manager, interview, 4 November 1993.

52 Ibid.

53 Ibid.

54 LU7689, minutes, 3 June 1991.

55 Ibid., 13 August 1991; 6 April 1992.

56 LU7689 President, interview, 6 October 1993.

57 See for example, Katz, *Shifting Gears*.

58 Atkinson, 'Manpower Strategies for Flexible Organisations'; Cox, *Production, Power and World Order*, ch. 9; D. Harvey, *The Condition of Postmodernity*.

59 *Globe and Mail*, Report on Business, Toronto Stock Exchange Quotations from 1991 to 1995.

60 Cameco holds 67 per cent of the assets in each of the Key Lake and Rabbit Lake mines. The remaining one-third is owned by the German concern, Uranerz. The other operating mine in Saskatchewan is 100 per cent owned by the French firm Cogema Resources.

61 Cameco Corp., *1994 Annual Report*.

62 Key Lake Mine Manager, interview, 10 January 1996.

63 Ibid.

64 USWA Staff Representative, interview, 6 September 1995.

65 On the subtle, but highly meaningful manifestations that power relations can assume see J. Scott, *Domination and the Arts of Resistance: Hidden Transcripts* (New Haven: Yale University Press, 1990).

66 Cameco, Key Lake, *Absenteeism Corrective Action Program*, 17 July 1995 (mimeograph).

67 Cameco, Key Lake, *Administering Discipline-Cameco Hourly Employees*, 19 January 1993 (mimeograph).

68 Of course, it is also the case that the institutions of civil society – household and community – no longer have the income earners available for what are considered to be normal periods of time.

69 Key Lake training foreman, interview, 9 January 1996. According to this individual, it is rare to have a full complement of workers on any given day. Also, Mine Manager, interview, 10 January 1996.

70 LU 8914 past President, interview, 10 May 1995.

71 Cameco, Key Lake, *Administering Discipline-Cameco Hourly Employees*, 19 January 1993.

72 Burawoy, *Politics of Production*, ch. 2 and 5; and 'The Functions and Reproduction of Migrant Labor.' Also see Wolpe, 'Capitalism and Cheap Labour-Power in South Africa'; C. van Onselen, *Chibaro: African Mine Labour in Southern Rhodesia, 1900–1933*.

73 Burawoy, *Politics of Production*, 92.

74 Ibid., 229.

75 Ibid., 97.

76 Ibid.

77 This is not to downplay the importance of colonialism as a key factor in northern Canadian history.

78 Respondent 325, interview, 31 January 1996.

79 LU8914 Bargaining Committee, meeting, 8 January 1996.

80 One of the electricians who was interviewed expressed an interest in more up to date training, especially in instrumentation technologies, which are now the preserve of special technicians. Respondent 325, interview, 31 January 1996. In the mining department, different workers expressed both a desire and the requisite

levels of past experience to operate pieces of equipment that they currently are forbidden to handle. This would make their jobs both more interesting and more remunerative. Respondents 407, 408, interview, 14 July 1995.

81 Respondent 325, interview, 31 January 1996.

82 Foreman in charge of mill training, interview, 9 January 1996.

Chapter 4: Corporate Cultures of Employment II

1 Noranda Inc., *Annual Report*, 1993.

2 Vigoro owns one other potash operation, also a Saskatchewan property called Kalium. This is one of the province's solution mines. G. Brock, 'Colonsay Potash Mine Purchase in Works,' *Star-Phoenix* (Saskatoon), 28 October 1994, B-8; A. Robertson, 'Noranda Sells Sask. Potash Mine,' *Globe and Mail*, 15 November 1994, B-5.

3 IMC Global, *Annual Report*, 1995; *IMC Global*, Press Release, 3 November 1995; C. Varcoe, 'Vigoro, IMC Global Share Swap Announced,' *Star-Phoenix*, 15 November 1995, C-8.

4 IMC Global Inc., *Annual Report*, 1995.

5 CCP Mine Manager, interview, 31 May 1995.

6 Local 7656 past Vice-President, interview, 5 February 1993; Respondent 218, interview, 3 November 1994; Local Union, minutes, 18 May 1982, 1 September 1982, 3 November 1982, and 11 April 1984; Memo, R.G. to Supervisors, Re: Arbitrator's Decision, 1981 Walkout, 20 September 1981. After three years of negotiations and litigation over the wildcat, the union was required to pay a fine of $5000 which was placed in trust into its own shop steward training fund!

7 C.C. Frewin, 'Management Issues at CCP,' March 1983 (mimeograph).

8 *Unit Reports*, Mine Department, and Mill Department, 10 March 1983.

9 C.C. Frewin, 'Management Issues at CCP.'

10 Mine Manager's Human Resource Statement, March 1984 (mimeograph).

11 LU7656 past Vice-President, interview, 5 February 1993; Local Union, minutes, 10 December 1984.

12 LU7656 past Vice-President, interview, 5 February 1993; Local Union, minutes, 10 December 1984.

13 Local Union, minutes, 10 July 1984, 16 September 1984, 10 December 1984, and 25 November 1985; Letter from LU7656 President to Mine Manager, 4 November 1985.

14 LU7656 past Vice-President, interview, 1 June 1995.

15 Pro-Active Consultants Ltd, *Central Canada Potash Change Project: Conclusions from Data Gathering and Revised Recommendations*, 25 January 1989 (mimeograph).

16 Respondents 774 nd 176, interviews, 25 October 1994; respondent 218, interview,

3 November 1994; Local Union, minutes, 22 November 1990, and monthly meeting, January 1991.

17 Meeting with Local Union 7656 Executive Committee, 2 June 1996.

18 Mine Manager, interview, 3 May 1995; Macro Consulting Group, 'Notes and Observations, Department Head Interviews,' 10 December 1991.

19 Local Union, minutes, 21 January 1988; LU7656 past Vice-President, interview, 1 June 1995.

20 With the change in ownership of CCP, this turns out to be more than an idle speculation. The new owners of the mine, IMC, have already made it known that performance sharing, or other like-minded programs, do not form part of its corporate philosophy.

21 Letter from USWA Canadian Office to T. S., re: performance-sharing system, 18 April 1988.

22 LU7656 President, interview, 14 December 1995.

23 Letter of Understanding Between CCP and USWA Local 7656, 31 May 1991. During the apprenticing period, seniority continued to accrue for the employee as if (s)he were still working in their old department.

24 Central Canada Potash, *Vision and Values Statement*, 1989.

25 Ibid., emphasis in the original.

26 Ibid., emphasis in the original.

27 It was noted by the consultants that this type of strategy had been successfully deployed in a number of organizations including the best known touchstone, Shell, Sarnia. Also see T. Rankin, *New Forms of Work Organization* and C. Heckscher, *The New Unionism*, for laudatory treatments of the Shell case.

28 Pro-Active Consultants, *Central Canada Potash Change Project: Conclusions from Data Gathering and Revised Recommendations*, 25 January 1989.

29 Ibid., 7.

30 Ibid.

31 For an insightful comparison of the earlier Employee Involvement and Quality of Working Life programs with the now fashionable Total Quality Management designs, see S. Hill, 'How Do You Manage a Flexible Firm? The Total Quality Model,' *Work, Employment and Society* 5, no. 3 (1991); and by the same author, 'Why Quality Circles Failed but Total Quality Management Might Succeed,' *British Journal of Industrial Relations* 29, no. 4 (1991).

32 Pro-Active Consultants, *Central Canada Potash Change Project*, 7.

33 Mine Manager, interview, 3 May 1995.

34 LU7656 past Vice-President, interview, 1 June 1995.

35 Mine Manager, interview, 3 May 1995.

36 Central Canada Potash, documentation on Employee Recognition Program, n.d. I have listed the criteria in the order in which they are weighted.

37 Innovation Team Report, 2 February 1994.

38 Mine Manager, interview, 3 May 1995.

39 Innovation Team Report, 2 February 1994.

40 See Katz, *Shifting Gears.*

41 Local 7656 President, interview, 17 October 1994.

42 Respondent 185, interview, 5 December 1994; respondent 226, 26 October 1994; Local Union, minutes, 10 January 1989, 25 March 1993, 15 June 1993, 31 March 1994.

43 Local Union, minutes, 10 March 1988.

44 Mine Manager, interview, 3 May 1995.

45 Respondent 176, interview, 25 October 1994.

46 Issues and Concerns of Supervisory Staff, January 1992, (mimeograph).

47 LU7656 President, interview, 5 December 1994; Local Union, minutes, 16 February 1990. Because the parallel apprenticeship program was agreed to in a letter of understanding between the local union and the company, the union has been unable to grieve the non-fulfilment of its provisions.

48 This possibility was raised during the course of an interview with a mill operator. Respondent 185, interview, 5 December 1994.

49 LU7656 President, interview, 5 December 1994; Local Union, minutes, 16 December 1993.

50 'Notes and observations, Department Head Interviews,' 10 December 1991. This was a follow-up audit to the 1989 climate survey that had been conducted at the site.

51 Cominco Ltd., Annual Report, 1991. The remaining 2.5 per cent of employment was located in other aspects of Cominco's business. These are corporate figures which include all employment on the company's payroll.

52 Henceforth I will use the company's current name Agrium when referring to it.

53 Cominco Fertilizers Ltd., *Annual Report*, 1993.

54 Notes from a meeting with the Mine Manager by the Union representative on the 4X Steering Committee, n.d.; Potash Competitiveness Team, *Employee Bulletin*, 30 September 1993.

55 I provide an initial analysis of this in B. Russell, 'Rival Paradigms at Work: Work Reorganization and Labour Force Impacts in a Staple Industry,' *Canadian Review of Anthropology and Sociology*, 34 no. 1 (1997).

56 LU 7552 past President, interview, 28 September 1993; LU 7552 past Vice-President, interview, 4 December 1993; Local Union, minutes, 17 May 1988.

57 LU 7552 past President, interview, 24 March 1994; respondent 81, interview, 20 June 1994; respondent 80, interview, 28 June 1994.

58 Mine Manager, interview, 1 November 1994.

59 LU7552 past Vice-President, interview, 4 December 1993.

60 'Cominco Fertilizers Goals and Objectives,' 11 May 1992; Mine Manager, interview, 1 November 1994.

61 'Cominco Fertilizers Goals and Objectives,' n.d., (mimeograph).

62 4X Joint Steering Committee, minutes, 3 April 1992; Participant's Notes on Using the Team Concept at Cominco, n.d.

63 The passage on teamwork is taken from the company's new mission statement which was written at the same time as the adoption of 4X.

64 Road to 4X Documentation Material, n.d.

65 LU7552 past Vice-President, interview, 14 December 1993; Mine Manager, interview, 1 November 1994. The local had been seriously interested in CWS, had put it on the negotiating table in 1980, and had struck a CWS committee in 1982 to promote the program. When the matter came to a vote, however, it was not carried. Apparently, a majority were convinced that they would have become locked into their current job positions, with less room for internal mobility than without CWS. Local Union, minutes, 18 January 1977, 15 March 1977, 19 June 1979, 16 December 1980, 11 May 1981, 20 October 1981, 21 December 1982, 18 January 1983, 18 February 1983, 15 March 1983, 19 April 1983.

66 Mine Manager, interview, 1 November 1994.

67 Management has costed labour expenses at Agrium out to be about $1.00 per hour more than the industry norm. Within our samples, the average job class positions at the other mines are certainly less than the journey operator rate which most workers receive at Agrium. Mine Manager, interview, 1 November 1994.

68 Road to 4X Documentation Material, n.d.

69 Ibid.

70 Cominco Fertilizers, *Mission Statement*, 1990.

71 4X Joint Steering Committee, minutes, 13 October, 20 October, 27 October 1992.

72 Company Statement on Interdependent, Self-Directed Work Teams, 25 November 1992.

73 Ibid.

74 Road to 4X Documentation Material, n.d.; Documentation on Employee Involvement Subcommittee, n.d.

75 R. Fisher and W. Ury, *Getting to Yes: Negotiating Agreement without Giving In* (New York: Penguin Books, 1991).

76 For example, Braverman, *Labor and Monopoly Capital*; M. Burawoy, *Manufacturing Consent: Changes in the Labor Process Under Monopoly Capitalism* (Chicago: University of Chicago Press, 1979) and *Politics of Production*; Edwards, *Contested Terrain*; D. Nelson, *Frederick W. Taylor and the Rise of Scientific Management*; and 'Scientific Management and the Workplace.'

77 4X Steering Committee, minutes, 21 May 1992.

78 4X Steering Committee, minutes of meeting to discuss job security, 8 September 1992; Job Security Subcommittee, minutes, 3 December 1992; Recognition Subcommittee, minutes, 8 December 1992.

79 Mine Manager, interview, 24 November 1994; 4X Steering Committee, minutes, 16 February 1993, 16 March 1993.

80 Addendum to Steering Committee, minutes, 17 March 1993.

81 4X Steering Committee, minutes, 30 April 1992, 6 January 1993, 13 January 1993, 20 January 1993, 16 February 1993.

82 Facilitators Meeting, minutes, 27 October 1993; 4X Steering Committee, minutes, 20 January 1993, 29 April 1993.

83 Mine Manager, memo, 10 March 1993.

84 Mine Manager, interview, 1 November 1994.

85 4X Joint Steering Committee, minutes, 9, 16 March 1993.

86 Letter of Understanding between USWA 7552 and Cominco Ltd., 11 September 1992.

87 Director of Human Resource Management, interview, 28 April 1994.

88 This is taken from 'The Executive Course in Continuous Improvement' that members of the steering committee at Agrium took. The seminar was designed by Tennessee Associates International Ltd, 1992.

89 Ibid., 5.

90 Ibid.

91 Ibid., 17–18.

92 Ibid., 7.

93 Under a continuous improvement program, each team authors its own charter, but the steering committee 'charters' each team. (Tennessee Associates, 1992).

94 Letter of Understanding, 11 September 1992.

95 Ibid.

96 Draft Copy of Proposed Guidelines for Continuous Improvement, 8 December 1993, 2.

97 Ibid., 5, capitalization in the original.

98 Ibid., capitalization in the original.

99 Ibid., 12.

100 Ibid., 25.

101 Ibid., 27.

102 Mine Manager, interview, 24 November 1994.

103 Draft Copy of Proposed Guidelines for Continuous Improvement, 32.

104 Ibid., 29; Draft Letter of Understanding, Potash Operations Gainshare Plan, 29 November 1994.

105 Cominco Fertilizers Ltd., Competency Development Program, 21 March 1995.

106 Competency Development Program, training session, 9 May 1995.

107 Ibid.

108 Remarks made at morning and afternoon sessions of the Competency Development Program, training session, 9 May 1995.

109 Agrium Training Coordinator, interview, 21 May 1995.

110 USWA, *Steelworker Guidelines for Participation in Work Reorganization*, 1992. As the document begins, 'Our union must be prepared to actively participate in work reorganization ... It is only when the union is actively involved that the positive potential of work reorganization will be fully developed and the negative consequences be minimized or eliminated.' 1.

111 During these week-long meetings, the local president took copious notes of members' largely negative responses to the program. This was then used as input for formulating a position on this program.

112 LU 7552 shop stewards' meeting, 16 January 1995.

113 Local Union past President, interview, 24 March 1994; Local Union President and USWA Staff Representative, interview, 1 July 1994.

114 Local Union, Special Meeting on 4X, minutes, 9 November 1991.

115 At one meeting a motion was put forward to withdraw from the program. At another, the film 'QWL, Management's Hidden Agenda' was screened. Local Union, minutes, 28 January 1992, 12 April 1992. Despite these ambivalences, the local did not drop its participation in the program at this stage.

116 Strategic Plan For Steelworker Local 7552, January 1993 (mimeograph).

117 Ibid.

118 Ibid., 18, 22.

119 Letter of Understanding Between Local 7552 and Cominco Ltd, 11 September 1992. The main objectives, as set out in the letter, included the creation of 'an environment where all employees feel secure in participating in processes that continuously improve Cominco Fertilizers' Potash Organization and Operation.' In order to accomplish this, the letter states that 'The union and its members and the company and its management staff are committed to participate in consensus process (solutions for mutual gain).'

120 Letter from USWA National Director regarding Union Strategic Plan, 23 June 1993; also USWA Steelworker Guidelines for Participation in Work Reorganization, 1992.

121 Letter from USWA Canadian Office to T.S. re. Performance-Sharing System, 18 April 1988. This correspondence originated around the Central Canada Potash profit-sharing scheme, but the policy advice would have been viewed as equally germane, and in some cases more so, for the Agrium plan.

122 With respect to this development, the Mine Manager suggested that the company had just become independent from the parent firm, Cominco metals, while at the same time it was facing an uncertain product market. Local management had failed to consider these factors before reaching the tentative agreement with the union. Mine Manager, interview, 1 November 1994; LU7552 past President, interview, 24 March 1994.

123 Local Union, Special Meeting on Continuous Improvement, minutes, 3 May 1993; Letter from Local Union President to Mine Manager, 1 September 1993.
124 In all, the company wanted to shed twelve hourly and twelve staff positions. LU7552 past President, interview, 17 February 1994; Respondent 105, interview, 10 March 1994.
125 LU7552 past President, interview, 7 April 1994.
126 Mine Manager, interview, 24 November 1994.
127 USWA Staff Representative, interview, 14 July 1994; Union study session, 21 July 1994.
128 Memorandum of Agreement, 22 June 1994.
129 Executive and shop stewards meeting, 8 July 1994; Bargaining committee meeting, 8 July 1994; Monthly membership meeting, 19 July 1994; Study session meetings, 21 July 1994.
130 LU7552 contract ratification meeting, 27 July 1994.
131 Mine Manager, interview, 1 November 1994.
132 Consultant's Report For Cominco Fertilizers, *Project 1, Gainsharing Plan Evaluation*, 23 September 1994.
133 Ibid., 8.
134 Ibid., 11, emphasis in the original.
135 Ibid., 13; also Mine Manager, interview, 1 November 1994.
136 Unsigned Potash Operations Gainshare Plan, Letter of Understanding, 29 November 1994.
137 Ibid.; and Local Union President, interview, 3 December 1994.
138 LU7552 shop stewards' meeting, 16 January 1995; Special membership meeting on company gainsharing plan, 17 January 1995.
139 Cominco Fertilizers, Employee Bulletin, 9 January 1995; Letter from Mine Manager to USWA Staff Representative, 6 January 1995.

Chapter 5: The Labour Process

1 The underground uranium mining that will be coming on stream in the next few years is another matter altogether. Ore bodies that are seven to eight times richer than existing mines render human exposure for even short periods of time infeasible. For these operations, underground mining will have to be fully automated with the use of robotic mining machines. We are currently standing on the edge of such developments, which represent the next generation of mining technology.
2 Cranes are used for various types of repair to mining equipment, such as rewelding. Although the crane operator is not strictly speaking involved in the mining of ore, the job incumbent is classified as being in the mining department.

3 Agrium site visit, 14 December 1993; Agrium Ltd, publication, 'Down to Earth People,' n.d.

4 See, for example, E. Trist et al., *Organizational Choice*, for a description of the much more complex division of labour in certain segments of the British coal mining industry.

5 PCS-Allan site visit, 23 April 1993; and LU7689 past President, interview, 23 April 1993.

6 The way desired production rates were expressed varied from mine to mine. At some, tonnes per hour or shift were mentioned, at others it was feet cut per hour or shift. This tended to depend on whether or not chevron cutting patterns were in use. Where they were, tonnage rates were usually volunteered by informants.

7 Agrium site visit, 14 December 1993; and LU7552 past Vice-President, interview, 14 December 1993.

8 PCS-Allan, site visit, 19 August 1992.

9 Respondent 176, interview, 25 October 1994; Local 7656, minutes, 22 November 1990, 22 February 1991, 26 April 1991, January 1992; and LU 7656, 'On the Local Scene' 1, no. 1 (May 1996).

10 In all of the potash sites under consideration the crushing facility (i.e. impaction area) is part of the mill complex, indeed, is the first labour process involved in milling. This would appear to be the industry norm.

11 This account is taken from an interview with the Key Lake training foreman, 9 January 1996.

12 Here, I am using the concepts of Taylorism and Fordism to refer strictly to the organization of the labour process, that is, to the utilization of the principles of scientific management plus moving assembly line forms of technology.

13 Hirschhorn, *Beyond Mechanization: Work and Technology in a Postindustrial Age*, 64, 66.

14 H. Braverman, *Labor and Monopoly Capital*.

15 D. Noble, *Forces of Production*; H. Shaiken, *Work Transformed: Automation and Labor in the Computer Age* (New York: Rinehart and Winston, 1984).

16 Littler, *The Development of the Labour Process in Capitalist Societies*; C. Littler, 'The Labour Process Debate,' in Knights and Willmott, eds., *Labour Process Theory* (London: Macmillan, 1990); Friedman, *Industry and Labour*; A. Friedman, 'Managerial Strategies, Activities, Techniques and Technology,' in *Labour Process Theory*.

17 J. Woodward, *Industrial Organization: Theory and Practice* (London: Oxford University Press, 1965).

18 Ibid., 63.

19 R. Blauner, *Alienation and Freedom* (Chicago: University of Chicago Press, 1964).

20 Ibid., 133.

21 Ibid., 134.

22 Ibid., 157.

23 Ibid., 143.

24 Ibid.

25 Hirschhorn, *Beyond Mechanization*, 27–31.

26 Zuboff, S. *In the Age of the Smart Machine* (New York: Basic Books, 1988), 10–12, 57; Hirschhorn, *Beyond Mechanization*, 91–8; also Block, *Postindustrial Possibilities: A Critique of Economic Discourse*, ch. 4.

27 Hirschhorn, *Beyond Mechanization*, 73.

28 P. Adler and B. Borys, 'Automation and Skill: Three Generations of Research on the NC Case,' *Politics and Society* 17, no. 3 (1989), 389; also P. Adler, 'New Technologies, New Skills,' *California Management Review* 29, no. 1 (1986).

29 Zuboff, *In the Age of the Smart Machine*, 40–1.

30 Ibid., 70–96.

31 Adler and Borys, 'Automation and Skill,' 386. The authors state that the use of computerized machine tools – numeric control – invalidates the Babbage principle, according to which work is consistently rationed more efficiently to productive tasks through the use of lower-waged, subdivided labour. Although the authors note higher-wage bills after the adoption of NC technology for given workforces, they do not control for the total size of the workforce.

32 Hirschhorn, *Beyond Mechanization*, 158.

33 To be fair some authors in this genre acknowledge and emphasize the determining role that managerial choice will still play, while deploring the incompatibilities between Taylorism and computerized technologies. One good example is Zuboff, *In the Age of the Smart Machine*.

34 Bell, *The Coming of Post-Industrial Society*.

35 Ibid., 137.

36 W. Clement and J. Myles, *Relations of Ruling: Class and Gender in Postindustrial Societies* (Montreal: McGill-Queen's University Press, 1994), 72; also J. Myles, 'The Expanding Middle: Some Canadian Evidence on the Deskilling Debate,' *Canadian Review of Sociology and Anthropology* 25, no. 3 (1988).

37 D. Gaullie, 'Patterns of Skill Change: Upskilling, Deskilling or the Polarization of Skills,' *Work, Employment and Society* 5, no. 3 (1991).

38 Indicators of skill, such as greater levels of in-house training may reflect the advent of more demanding (i.e., skilled) jobs or, then again, they may be largely directed at resocializing workers into new managerial cultures, as in the case of Agrium's competency development program.

39 Braverman lent priority to the accumulation of capital as the overriding purpose of modern production, from which all else flowed. Technology, in and of itself, was a neutral factor, overdetermined by the social purposes of labour. For Noble, on the

other hand, control of the labour process reigned supreme. In this quest, technologies were specifically designed to enhance managerial control, a point with which Shaiken concurs. Braverman, *Labor and Monopoly Capital*; Noble, *Forces of Production*; Shaiken, *Work Transformed.*

40 Braverman, *Labor and Monopoly Capital*, 81–2.

41 Ibid., p. 180. Significantly, Braverman notes that this is more a description of business nirvana than a depiction of contemporary reality. Nonetheless, and just as important, it remains a desideratum of every employer.

42 Ibid., 85, 225, 307–25.

43 Ibid., 329–30, 338.

44 This, for example, is the position of Zuboff, who argues that the new skills sets occasioned by informating technologies are so different from the tactile skills of yesteryear as to be non-comparable. See Zuboff, *In the Age of the Smart Machine.*

45 Braverman, *Labor and Monopoly Capital*, 426–35. On this point Braverman notes that, 'The creation of "semi-skill" ... thus brought into existence, retroactively to the turn of the century and with a mere stroke of the pen, a massive "upgrading" of the skills of the working population. By making a connection with machinery ... a criterion of skill, it guaranteed that with the increasing mechanization of industry the category of the "unskilled" would register a precipitous decline, while that of the "semi-skilled" would show an equally striking rise ... The entire concept of 'semi-skill,' as applied to operatives, is an increasingly delusory one ... for the category of operatives, training requirements and the demands of the job are now so low that one can hardly imagine jobs that lie significantly below them on any scale of skill.'

46 Ibid., 26–30.

47 Burawoy, *Politics of Production*, 35–40.

48 See for example J. Gaskell, 'What Counts as Skill? Reflections on Pay Equity' in *Just Wages: A Feminist Assessment of Pay Equity*, ed. J. Fudge and P. McDermott (Toronto: University of Toronto Press, 1991); J. Jenson, 'The Talents of Women, the Skills of Men: Flexible Specialization and Women,' in S. Wood, ed., *The Transformation of Work?* (London: Unwin Hyman, 1989).

49 J. Bright, 'Does Automation Raise Skill Requirements?' *Harvard Business Review* (1958), 87.

50 Ibid., 87.

51 Ibid., 90–5.

52 K. Dix, 'Work Relations in the Coal Industry: The Handloading Era, 1880–1930,' in A. Zimbalist, ed., *Case Studies on the Labor Process* (New York: Monthly Review Press, 1979), 163.

53 Ibid., 167–9.

54 M. Yarrow, 'The Labor Process in Coal Mining,' in *Case Studies on the Labor Process*.

55 Ibid., 185.

56 Ibid., 191.

57 Clement, *Hardrock Mining: Industrial Relations and Technological Changes at Inco*.

58 Ibid., 161.

59 Ibid., 163.

60 Ibid., 217–18.

61 Specifically, Clement addresses the alternative of responsible autonomy as advanced by Friedman. For Clement there is little evidence that such strategies 'are either widespread or persistent. There is more evidence from the mining industry that managers turn to mechanization or automation rather than accommodate hostile workers.' Ibid., 254; Friedman, *Industry and Labour*.

62 Hall, 'Production Politics and the Construction of Consent: The Case of Health and Safety in Mining,' 371–2, 375, 383, 389.

63 Ibid., 376–81, 435–6, 540–2.

64 E. Trist and K. Bamforth, 'Some Social and Psychological Consequences of the Longwall Method of Coal Getting,' *Human Relations* 4, no. 1 (1951); J. Goldthorpe, 'Technical Organization As A Factor in Supervisor-Worker Conflict,' *British Journal of Sociology* 10 (1959); E. Trist et al., *Organizational Choice*.

65 Trist and Bamforth, 'Some Social and Psychological Consequences,' 23; Trist et al., *Organizational Choice*, 46.

66 Trist et al., *Organizational Choice*, 143.

67 Ibid., 78, 86–7, 242, 247–8, 254.

68 Friedman, *Industry and Labour*, 6.

69 Ibid., 99.

70 Ibid., 82–4, 101; Friedman, 'Managerial Strategies, Activities, Techniques and Technology.'

71 Rankin, *New Forms of Work Organization*, 5, 30. This observation seems quite at odds with Trist's own position that with the development of mechanization 'no longer is the producer a man serviced by machines but a machine serviced by men. For the possibilities of this new situation to be realized a change is required in the work culture from a man-centered to a machine-centered attitude – a machine culture.' Trist et al., *Organizational Choice*, 259. Dohse et al. also note the decentralization of responsibility that is associated with Japanese production relations when they contrast 'the Japanese "responsibility model" ... to the "control model"' of Fordist labour processes. K. Dohse, et al. 'From "Fordism" to "Toyotism"?' 125.

72 Hirschhorn, *Beyond Mechanization*, 115–17; Rankin, *New Forms of Work Organization*, 69–71.

73 J. Womack et al., *The Machine That Changed the World*. A thoughtful challenge to these claims is contained in K. Williams et al., 'Against Lean Production,' *Economy and Society* 21 no. 3 (1992).

74 Womack et al., *The Machine That Changed the World*, 13–14. 'Lean production calls for learning far more professional skills and applying these creatively in a team setting rather than in a rigid hierarchy.'

75 Ibid., 102.

76 Kenney and Florida, *Beyond Mass Production*.

77 Ibid., 68.

78 Ibid., 15, 124, 271.

79 Ibid., 15, and 54, 74. This is similar to the new skill profiles that Rankin puts forward, which include assuming responsibility for heavily capital intensive technologies, a willingness to engage in abstract reasoning, and deployment of the social skills required to work in a team environment. Rankin, *New Forms of Work Organization*, 54.

80 Kenney and Florida, *Beyond Mass Production*, 66, 304.

81 J. Price, 'Lean Production at Suzuki and Toyota: A Historical Perspective,' 87.

82 Kenney and Florida, *Beyond Mass Production*, 38.

83 Burawoy, *Politics of Production*, 127–8, 148–52.

84 K. Dohse, 'From "Fordism" to "Toyotism"?' 141.

85 J. Price, 'Lean Production at Suzuki and Toyota.'

86 S. Wood, ed. *The Transformation of Work?* 20, 34.

87 C. Berggren, *Alternatives to Lean Production*; also see J. Fucini and S. Fucini, *Working for the Japanese: Inside Mazda's American Auto Plant* (New York: Free Press, 1990).

88 Berggren, *Alternatives to Lean Production*, 5.

89 Ibid., 29.

90 Ibid., 30.

91 Ibid., 43–4.

92 D. Robertson et al., 'Team Concept and Kaizen: Japanese Production Management in a Unionized Canadian Auto Plant,' *Studies in Political Economy* 39 (1992). Also see R. Milkman, *Farewell to the Factory*, Chapter 5, where post-Fordist work relations in the American automobile industry are equated with the further deskilling of assembly work.

93 M. Parker and J. Slaughter, *Choosing Sides: Unions and the Team Concept* (Boston: South End Press, 1988).

94 Ibid., 80.

95 M. Kenney and R. Florida, *Beyond Mass Production*, 110.

96 Ibid., 133.

97 Ibid., 135.

98 J. Corman et al., *Recasting Steel Labour*, 39, 49–50.

99 Ibid., 15, 148–9.

100 In this and the cross-tabulation tables which follow, raw frequencies are given first, followed by the most relevant proportion. The appropriate measure of statistical association and the level of significance are also provided.

101 Gaullie, 'Patterns of Skill Change.'

102 At individual worksites these same patterns are more or less repeated. That is, at all of the complexes, with the exception of Central Canada Potash, mill operators ranked either first or second amongst the occupational groups, in terms of registering increasing skill levels. The results carried high levels of statistical association between occupation and reported skill trend and were statistically significant at three of the mine sites. For PCS-Allan, Agrium, and Cameco the results were statistically significant at the .05 level or less, and carried high levels of association with them (Cramer's V = .32 to .39). The findings for Central Canada and PCS-Cory were not statistically significant.

103 Hirschhorn, *Beyond Mechanization*, 67, 123; Zuboff, *In the Age of the Smart Machine*.

Chapter 6: Production Politics at Five Mine Sites

1 H.C. Pentland, *A Study of the Changing Social, Economic, and Political Background of the Canadian System of Industrial Relations* (Draft study prepared for the Task Force on Labour Relations, 1968, mimeograph), 1.

2 Ibid.

3 H. Clegg, *The Changing System of Industrial Relations in Great Britain* (Oxford: Basil Blackwell, 1983), 1.

4 M. Burawoy, *The Politics of Production*, 122.

5 See for example, Russell, *Back to Work? Labour, State and Industrial Relations in Canada*; Dubofsky, *The State and Labor in Modern America*; Tomlins, *The State and the Unions*; and Edwards, *Conflict at Work* for works that deal with Canada, the United States, and Britain respectively.

6 Rankin, *New Forms of Work Organization*, 20; also, Kochan et al., *The Transformation of American Industrial Relations*, 239–40.

7 M. Piore and C. Sabel, *The Second Industrial Divide*, 112–15.

8 Kochan et al., *The Transformation of American Industrial Relations*, 29.

9 H. Katz, *Shifting Gears*, 40–1.

10 Rankin, *New Forms of Work Organization*, 35.

11 For example, Womach et al., *The Machine That Changed the World*, 14, 99–102, 225; Kenney and Florida, *Beyond Mass Production*, 15–16, 27, 38, 54 and *passim*.

12 See Rankin, *New Forms of Work Organization*, 70 and *passim*.

13 Katz, *Shifting Gears*, 102.

14 Rankin reports a very low incidence of grievances in his study of Shell Sarnia, while Kochan et al. found QWL plans to have a negligible effect on grievances or absenteeism. See Rankin, *New Forms of Work Organization*, 99; and Kochan et al. *Transformation*, 156.

15 Price, 'Lean Production at Suzuki and Toyota,' 90.

16 See for example, Fucini and Fucini, *Working for the Japanese*, ch. 6.

17 See for example, Parker and Slaughter, *Choosing Sides*, 19; and Berggren, *Alternatives to Lean Production*, 30–2.

18 See for example, J. Rinehart, 'Appropriating Worker's Knowledge: Quality Control Circles at a General Motors Plant,' *Studies in Political Economy* 14 (1984); D. Noble, *Forces of Production*, ch. 11; D. Wells, *Empty Promises: Quality of Work Life Programs and the Labor Movement*. See also Robertson, et al., 'Team Concept and Kaizen.'

19 Wells, *Empty Promises*, 89–92. This point is also recognized as a possibility by advocates of team approaches, although they are outweighed by the possibilities for democratization that such reforms open up for Katz, *Shifting Gears*, 99; and for Rankin, *New Forms of Work Organization*, 129–30.

20 Wells, *Empty Promises*, 87.

21 Berggren, *Alternatives to Lean Production*, 228.

22 Ibid., 6.

23 Wells, *Empty Promises*, 5.

24 This is the case for both Agrium and Key Lake. At the other mines there is no consequential relationship between the issue involved in disciplinary cases and the filing of a grievance.

25 Haiven, 'Hegemony and the Workplace: The Role of Arbitration,' 105.

26 Grievances signed in support of, or on behalf of other workers are omitted from these calculations.

27 On this point, see Scott, *Hidden Transcripts of Resistance*.

28 Edwards and Scullion, *The Social Organization of Industrial Conflict*, 154.

29 L. Taylor and P. Walton, 'Industrial Sabotage: Motives and Meanings,' in *Images of Deviance,* ed. S. Cohen (Harmondsworth: Penguin Books, 1971), 219.

30 The statistical association (Cramer's V) between vandalism and mine site was .28 at a level of significance of .001.

31 These are the type of findings that are reported upon by Witte. J. Witte, *Democracy, Authority and Alienation in Work* (Chicago: University of Chicago Press, 1980).

32 Generally blasting is used in the underground mines under consideration here only to relieve blockages in storage bins. It is not a normal part of the actual mining activity. In other potash mines, however, such as one that I visited in the United Kingdom, blasting is used as part of the ore recovery process. First the rock face is

set with charges, which are exploded to release built-up gas pressure. Only then can cutting operations with a miner commence. Blasting does remain a regular procedure in open-pit uranium mining.

33 On this point also see Clement, *Hardrock Mining: Industrial Relations and Technological Changes at Inco*, 223; Hall, 'Production Politics and the Construction of Consent: The Case of Health and Safety in Mining,' 301, 329, 583.

34 Central Canada Potash was the only mine where a majority of those experiencing health problems did not attribute them to their work. Still, 46 per cent of the workers at this mine who indicated suffering from a chronic problem related it to the work environment.

35 A useful introduction to this topic is to be found in G. Lowe and H. Northcott, *Under Pressure: A Study of Job Stress* (Toronto: Garamond, 1986).

36 See for example, Parker and Slaughter, *Choosing Sides*; Berggren, *Beyond Lean Production*; and Fucini and Fucini, *Working for the Japanese*. Parker and Slaughter coined the term, 'management by stress.'

37 See for example, C. Cuneo, 'Trade Union Leadership: Sexism and Affirmative Action,' in *Women Challenging Unions: Feminism, Democracy and Militancy*, ed. L. Briskin and P. McDermott (Toronto: University of Toronto Press, 1993).

38 Also see T. Dunk, *It's a Working Man's Town: Male Working Class Culture in Northwestern Ontario* (Montreal: McGill-Queen's University Press, 1991) for a valuable analysis of white male working-class culture in a staples producing milieu.

Chapter 7: Final Reflections

1 Memorandum of Agreement, Cominco Fertilizers and USWA7552, 22 July 1994, (mimeograph); Agreement, Cominco Ltd. and USWA7552, 1 May 1991.

2 USWA 7552, Local Union History and Clippings Book.

3 Between Central Canada and PCS there was a $0.12 per hour differential.

4 S. Burnett, 'Agrium Potash Mine Workers Walk out over Wages, Pensions,' *Star-Phoenix* (Saskatoon), 13 May 1996, A-3.

5 As reported at the shop stewards' meeting, USWA7552, 9 May 1996.

6 It was claimed that Agrium was receiving from seventy to seventy-five cars of potash per week from a neighbouring mine to allow it to fulfill its contractual obligations. USWA7552 Ratification meeting, 3 September 1996.

7 It is unclear whether a petition for a vote on the company's last offer had been started or was merely being threatened. Information for this section was gathered at the two ratification meetings which the local held on 3 September 1996.

8 This was evident from remarks made by both the National Director and the District Director at the District 3 USWA Conference, 18–20 April 1996.

9 J. Parker, 'Potash Mine Unions Propose United Front,' *Star-Phoenix,* 8 May 1996, A-3.
10 Two questions in the survey dealt with the question of working hours. Only 20 per cent of those who had recently experienced lay-off were in favour of reduced working hours as a solution to the problem. Only 14 per cent of the total sample stated that more vacation time was a priority in future rounds of collective bargaining.

References

Adler, P., 'New Technologies, New Skills.' *California Management Review* 29, no. 1 (1986): 9–28.

Adler, P., and B. Borys. 'Automation and Skill: Three Generations of Research on the NC Case.' *Politics and Society* 17, no. 3 (1989): 377–402.

Aglietta, A. *A Theory of Capitalist Regulation.* London: New Left Books, 1979.

Arrighi, G. *The Long Twentieth Century.* London: Verso, 1995.

Atkinson, J. 'Manpower Strategies for Flexible Organisations.' *Personnel Management*, (1984): 28–31.

Baglioni, G., and C. Crouch, eds. *European Industrial Relations: The Challenge of Flexibility.* London: Sage, 1990.

Baldamus, W. *Efficiency and Effort.* London: Tavistock, 1961.

Bean, R. 'The "Cooperative Wage Study" and the Canadian Steelworkers.' *Relations Industrielles* 19, no. 1 (1964): 55–69.

Bell, D. *The Coming of Post-Industrial Society.* New York: Basic Books, 1973.

Berggren, C. *Alternatives to Lean Production.* Ithaca, NY: ILR Press, 1992.

Bird, F. *Mine Safety and Loss Control Management.* Loganville, GA: Institute Press, 1980.

Bird, F., and R. Loftus. *Loss Control Management.* Atlanta: Institute Press, 1976.

Blauner, R. *Alienation and Freedom.* Chicago: University of Chicago Press, 1964.

Block, F. *The Origins of International Economic Disorder.* Berkeley: University of California Press, 1977.

– *Post-Industrial Possibilities: A Critique of Economic Discourse.* Berkeley: University of California Press, 1990.

Bobiash, D. 'Attempts to Promote Industrial Democracy in the Potash Corporation of Saskatchewan,' MSc research paper in Industrial Relations and Personnel Management, London School of Economics, 1984.

Boyer, R. *The Regulation School: A Critical Introduction.* New York: Columbia University Press, 1990.

Braverman, H. *Labor and Monopoly Capital.* New York: Monthly Review Press, 1974.

Bright, J. 'Does Automation Raise Skill Requirements?' *Harvard Business Review* 36, no. 4 (1958): 85–98.

Brody, D. *Workers in Industrial America.* New York: Oxford University Press, 1980.

Burawoy, M. 'The Functions and Reproduction of Migrant Labor: Comparative Material from South Africa and the United States,' *American Journal of Sociology* 81, no. 5 (1976): 1050–87.

– *Manufacturing Consent: Changes in the Labor Process under Monopoly Capitalism.* Chicago: University of Chicago Press, 1979.

– *Politics of Production.* London: Verso, 1985.

Cameco Corporation. *Annual Reports* (various years).

Cameco Corporation and United Steel Workers of America Local 8914. *Collective Agreement.* 1993.

Canada. *Report of the Royal Commission on Dominion–Provincial Relations.* Books 1 and 2. Ottawa: King's Printer, 1940.

– Minister of Reconstruction, *Employment and Income.* Ottawa: King's Printer, 1945.

– Natural Resources Canada. *Canadian Minerals Yearbook.* Ottawa (various years).

– *Manufacturing Industries of Canada: National and Provincial Areas.* Ottawa: Statistics Canada, 1993.

Carroll, W. 'The Canadian Corporate Elite: Financiers or Finance Capitalists?' *Studies in Political Economy* 8 (1982): 89–114.

– *Corporate Power and Canadian Capitalism.* Vancouver: UBC Press, 1986.

Clawson, D. *Bureaucracy and the Labor Process.* New York: Monthly Review Press, 1980.

Clegg, H. *The Changing System of Industrial Relations in Great Britain.* Oxford: Basil Blackwell, 1983.

Clement, W. *Continental Corporate Power.* Toronto: McClelland and Stewart, 1977.

– *Hardrock Mining: Industrial Relations and Technological Changes at Inco.* Toronto: McClelland and Stewart, 1981.

– 'Debates and Directions: A Political Economy of Resources.' In W. Clement and G. Williams (eds.), *The New Canadian Political Economy,* 36–53. Montreal: McGill-Queen's University Press, 1989.

Clement, W., and J. Myles. *Relations of Ruling: Class and Gender in Postindustrial Societies.* Montreal: McGill-Queen's University Press, 1994.

Coates, D. *The Crisis of Labour.* Oxford: Philip Allan, 1989.

Cohen, S., and J. Zysman. *Manufacturing Matters: The Myth of the Post-Industrial Economy.* New York: Basic Books, 1987.

Cominco Fertilizers Ltd. *Annual Report.* 1993.

Cominco Ltd. *Annual Report*. 1991.

Corman, J. 'The Impact of State Ownership on a State Proprietary Corporation: The Potash Corporation of Saskatchewan.' PhD dissertation, University of Toronto, 1982.

Corman, J., et al. *Recasting Steel Labour*. Halifax: Fernwood, 1993.

Cox, R. *Production, Power, and World Order: Social Forces in the Making of History*. New York: Columbia University Press, 1987.

Cuneo, C. 'Trade Union Leadership: Sexism and Affirmative Action.' In L. Briskin and P. McDermott (eds.), *Women Challenging Unions: Feminism, Democracy, and Militancy*, 109–36. Toronto: University of Toronto Press, 1993.

Denis, N., et al. *Coal Is Our Life*. London: Eyre and Spottiswoode, 1956.

Dix, K. 'Work Relations in the Coal Industry: The Handloading Era, 1880–1930.' In A. Zimbalist (ed.), *Case Studies on the Labor Process*, 156–69. New York: Monthly Review Press, 1979.

Dohse, K., et al. 'From "Fordism" to "Toyotism": The Social Organization of the Labor Process in the Japanese Automobile Industry.' *Politics and Society* 14, no. 2 (1985): 115–46.

Drache, D. 'The Way Ahead for Ontario.' In D. Drache (ed.), *Getting On Track: Social Democratic Strategies For Ontario*, 217–37. Montreal: McGill-Queen's University Press, 1992.

Drache, D., and W. Clement. *The New Practical Guide to Canadian Political Economy*. Toronto: James Lorimer, 1985.

Drache, D. and H. Glasbeek. 'The New Fordism in Canada: Capital's Offensive, Labour's Opportunity.' *Osgoode Hall Law Journal*, 27, no. 3 (1989): 517–60.

– *The Changing Workplace: Reshaping Canada's Industrial Relations System*. Toronto: James Lorimer, 1992.

Dubofsky, M. *The State and Labor in Modern America*. Chapel Hill: University of North Carolina Press, 1994.

Dunk, T. *It's a Working Man's Town: Male Working-Class Culture in Northwestern Ontario*. Montreal: McGill-Queen's University Press, 1991.

Dunk, T., S. McBride, and R. Nelson, eds. *The Training Trap: Ideology, Training, and the Labour Market*. Halifax: Fernwood, 1996.

Edwards, P. *Conflict at Work*. Oxford: Basil Blackwell, 1986.

– *Managing the Factory*. Oxford: Basil Blackwell, 1987.

Edwards, P., and H. Scullion. *The Social Organization of Industrial Conflict*. Oxford: Basil Blackwell, 1982.

Edwards, R. *Contested Terrain*. New York: Basic Books, 1979.

Esping-Andersen, G. *The Three Worlds of Welfare Capitalism*. Princeton: Princeton University Press, 1990.

Ferner, A., and R. Hyman, eds. *Industrial Relations in the New Europe*. Oxford: Basil Blackwell, 1992.

Fisher, R., and W. Ury. *Getting to Yes: Negotiating Agreement without Giving in.* New York: Penguin Books, 1991.

Fowke, V. *The National Policy and the Wheat Economy.* Toronto: University of Toronto Press, 1978.

Freedman, R., and J. Medoff. *What Do Unions Do?* New York: Basic Books, 1984.

Friedman, A. *Industry and Labour.* London: Macmillan, 1977.

– 'Managerial Strategies, Activities, Techniques, and Technology.' In D. Knights and H. Willmott (eds.), *Labour Process Theory,* 177–208. London: Macmillan, 1990.

Fucini, J., and S. Fucini. *Working for the Japanese: Inside Mazda's American Auto Plant.* New York: Free Press, 1990.

Fudge, J. 'Voluntarism and Compulsion: The Canadian Federal Government's Intervention in Collective Bargaining from 1900 to 1946.' PhD dissertation, University College, Oxford University, 1987.

Gamble, A. *The Free Economy and the Strong State: The Politics of Thatcherism.* London: Macmillan, 1988.

Gaskell, J. 'What Counts as Skill? Reflections on Pay Equiaty.' In J. Fudge and P. McDermott (eds.), *Just Wages: A Feminist Assessment of Pay Equity,* 141–59. Toronto: University of Toronto Press, 1991.

Gaullie, D. 'Patterns of Skill Change: Upskilling, Deskilling or the Polarization of Skills.' *Work, Employment and Society* 5, no. 3 (1991): 319–51.

Giangrande, C. *The Nuclear North: The People, the Regions, and the Arms Race.* Toronto: Anansi, 1983.

Glasbeek, H. 'Labour Relations Policy and Law as Mechanisms of Adjustment.' *Osgoode Hall Law Journal* 25, no. 1 (1987) 179–237.

Goldthorpe, J. 'Technical Organization as a Factor in Supervisor-Worker Conflict.' *British Journal of Sociology,* 10, no. 3 (1959): 213–29.

Gonick et al., eds. *Labour Pains, Labour Gains: 50 Years of PC1003.* Halifax: Fernwood, 1995.

Gouldner, A. *Patterns of Industrial Bureaucracy.* New York: Free Press, 1964.

– *Wildcat Strike.* Yellow Springs, OH: Antioch Press, 1964.

Gramsci, A. *Selections from the Prison Notebooks.* London: Lawrence and Wishart, 1971.

Gullickson, D. 'Uranium Mining, the State and Public Policy in Saskatchewan, 1971–1982: The Limits of the Social Democratic Imagination.' MA thesis, University of Regina, 1990.

Haiven, L. 'The Political Apparatuses of Production: Generation and Resolution of Industrial Conflict in Canada and Britain.' PhD dissertation, University of Warwick 1988.

– 'Hegemony and the Workplace: The Role of Arbitration.' In S. McBride et al. (eds.), *Regulating Labour,* 79–117. Toronto: Garamond Press, 1991.

Hall, A. 'Production Politics and the Construction of Consent: The Case of Health and Safety in Mining.' PhD dissertation, University of Toronto, 1989.

– 'The Corporate Construction of Occupational Health and Safety: A Labour Process Analysis.' *Canadian Journal of Sociology* 18, no. 1 (1993): 1–20.

Harding, B. 'The Two Faces of Public Ownership: From the Regina Manifesto to Uranium Mining.' In J. Harding (ed.), *Social Policy and Social Justice: The NDP Government in Saskatchewan during the Blakeney Years*, 281–313. Waterloo: Wilfrid Laurier Press, 1995.

Harris, H. *The Right to Manage.* Madison: University of Wisconsin Press, 1982.

Harrison, B., and B. Bluestone. *The Great U-Turn: Corporate Restructuring and the Polarizing of America.* New York: Basic Books, 1988.

Harvey, D. *The Condition of Postmodernity.* Oxford: Basil Blackwell, 1989.

Heckscher, C. *The New Unionism.* New York: Basic Books, 1988.

Hill, S. 'How Do You Manage a Flexible Firm?' *Work, Employment and Society* 5, no. 3 (1991): 397–415.

– 'Why Quality Circles Failed but Total Quality Management Might Succeed.' *British Journal of Industrial Relations* 29, no. 4 (1991): 541–68.

Hirschhorn, L. *Beyond Mechanization: Work and Technology in a Post-Industrial Age.* Cambridge, MA: MIT Press, 1984.

IMC Global. *Annual Report*, 1995.

Innis, H. *The Fur Trade in Canada: An Introduction to Canadian Economic History.* Toronto: University of Toronto Press, 1977.

International Loss Control Institute. 'Mine Safety and Loss Control.' Loganville, GA: 1984. Mimeo.

Jacoby, S. *Employing Bureaucracy.* New York: Columbia University Press, 1985.

Jaikumar, R. 'Postindustrial Manufacturing.' *Harvard Business Review*, 6 (1986): 69–76.

Jenson, J. '"Different" but Not "Exceptional": Canada's Permeable Fordism.' *Canadian Review of Sociology and Anthropology* 26, no. 1 (1989): 69–94.

– 'The Talents of Women, the Skills of Men: Flexible Specialization and Women.' In S. Wood (ed.), *The Transformation of Work*, 141–55. London: Unwin Hyman, 1989.

Jenson, J. and R. Mahon, 'Legacies for Canadian Labour of Two Decades of Crisis.' In J. Jenson and R. Mahon (eds.), *The Challenge of Restructuring: North American Labor Movements Respond*, 72–92. Philadelphia: Temple University Press, 1993.

– 'North American Labour: Divergent Trajectories.' In J. Jenson and R. Mahon (eds.), *The Challenge of Restructuring: North American Labor Movements Respond*, 3–15. Philadelphia: Temple University Press, 1993.

Jessop, B. *State Theory.* Oxford: Polity Press, 1990.

Jessop, B., et al. *Thatcherism.* Oxford: Polity Press, 1988.

Johnson, A., S. McBride, and P. Smith (eds.). *Continuities and Discontinuities: The*

Political Economy of Social Welfare and Labour Market Policy in Canada. Toronto: University of Toronto Press, 1994.

Katz, H. *Shifting Gears.* Cambridge, MA: MIT Press, 1985.

Kelly, J. *Trade Unions and Socialist Politics.* London: Verso, 1988.

Kenney, M., and R. Florida. 'Beyond Mass Production: Production and the Labor Process in Japan.' *Politics and Society* 16, no. 1 (1988): 121–58.

– *Beyond Mass Production.* New York: Oxford University Press, 1993.

Keynes, J.M. *Essays In Persuasion.* London: Macmillan, 1953.

Klare, K. 'Judicial Deradicalization of the Wagner Act and the Origins of Modern Legal Consciousness.' *Minnesota Law Review* 62, no. 3 (1978): 265–339.

Knights, D., and H. Willmott. *Labour Process Theory.* London: Macmillan, 1990.

Kochan, T., H. Katz, and R. McKersie. *The Transformation of American Industrial Relations.* New York: Basic Books, 1986.

Laux, J.K., and M.A. Molot. *State Capitalism: Public Enterprise in Canada.* Ithaca: Cornell University Press, 1988.

Lazonick. W. *Business Organization and the Myth of the Market Economy.* Cambridge: Cambridge University Press, 1991.

Lichtenstein, N. *Labor's War at Home.* Cambridge: Cambridge University Press, 1982.

Lipietz, A. *The Enchanted World.* London: Verso, 1985.

– *Mirages and Miracles.* London: Verso, 1987.

Littler, C. *The Development of the Labour Process in Capitalist Societies.* London: Heinemann, 1982.

– 'The Labour Process Debate.' In D. Knights and H. Willmott (eds.), *Labour Process Theory*, 46–94. London: Macmillan, 1990.

Livingstone, D. 'Post-Industrial Dynamics: Economic Restructuring, Underemployment and Lifelong Learning in the Information Age.' Twenty-seventh Annual Sorokin Lecture, University of Saskatchewan, 1996. Mimeo.

Lowe, G., and H. Northcott. *Under Pressure: A Study of Job Stress.* Toronto: Garamond Press, 1986.

Luxton, M. *More Than a Labour of Love.* Toronto: Women's Press, 1980.

Luxton, M., and J. Corman. 'Getting to Work: The Challenge of the Women Back into Stelco Campaign.' *Labour/Le Travail* 28 (1991): 149–85.

MacInnes, J. *Thatcherism at Work.* Milton Keynes: Open University Press, 1987.

Mahon, R. 'From Fordism To? New Technology, Labour Markets and Unions.' *Economic and Industrial Democracy* 8, no. 1 (1987): 5–60.

– 'Post-Fordism: Some Issues for Labour.' In D. Drache and M. Gertler (eds.), *The New Era of Global Competition. State Policy and Market Power*, 316–32. Montreal: McGill-Queen's University Press, 1991.

Marchack, P. *Green Gold: The Forest Industry in British Columbia.* Vancouver: UBC Press, 1983.

Marx, K. *Capital*. Vol. 3. Moscow: Progress Publishers, 1971.

McBride, S. 'Hard Times and "The Rules of the Game."' In R. Argue et al. (eds.), *Working People and Hard Times*, 98–111. Toronto: Garamond, 1987.

– *Not Working: State, Unemployment, and Neo-Conservatism in Canada*. Toronto: University of Toronto Press, 1992.

McBride, S., and J. Shields. *Dismantling a Nation*. Halifax: Fernwood, 1993.

Milkman, R. *Farewell to the Factory*. Berkeley: University of California, 1997.

Millar, D. 'The Shapes of Power: The Ontario Labour Relations Board, 1944 to 1950.' PhD dissertation, York University, 1980.

Molot, M., and J. Laux. 'The Politics of Nationalization.' *Canadian Journal of Political Science* 12, no. 2 (1979): 227–58.

Montgomery, D. *Worker's Control in America*. Cambridge: Cambridge University Press, 1980.

Murray, M., ed. *South African Capitalism and Black Political Opposition*. Cambridge, Ma: Schenkman Publishing, 1982.

Myles, J. 'The Expanding Middle: Some Canadian Evidence on the Deskilling Debate.' *Canadian Review of Sociology and Anthropology* 25, no. 3 (1988): 335–64.

Naylor, T. *The History of Canadian Business, 1867–1914*. 2 vols. Toronto: Lorimer, 1975.

Nelson, D. *Fredrick W. Taylor and the Rise of Scientific Management*. Madison: University of Wisconsin Press, 1980.

– 'Scientific Management and the Workplace.' In S. Jacoby (ed.), *Masters to Managers*, 74–89. New York: Columbia University Press, 1991.

Niosi, J. *Canadian Capitalism: A Study of Power in the Canadian Business Establishment*. Toronto: Lorimer, 1981.

Noble, D. *Forces of Production*. New York: Oxford University Press, 1986.

– *Progress without People*. Toronto: Between the Lines Press, 1995.

Noranda Inc. *Annual Report*, 1993.

Olsen, G. 'Locating the Canadian Welfare State.' *Canadian Journal of Sociology* 19, no. 1 (1994): 1–20.

van Onselen, C. *Chibaro: African Mine Labour in Southern Rhodesia, 1900–1933*. London: Pluto Press, 1976.

Panitch, L. 'Dependency and Class in Canadian Political Economy.' *Studies in Political Economy* 6 (1981): 7–33.

Panitch, L., and D. Swartz. 'Towards Permanent Exceptionalism.' *Labour/Le Travail*, 13 (1984): 133–57.

– *From Consent to Coercion*. Toronto: Garamond, 1985.

– *The Assault on Trade Union Freedoms*. Toronto: Garamond, 1988.

– *The Assault on Trade Union Freedoms: From Wage Controls to Social Contract*. Toronto: Garamond, 1993.

Parboni, R. *The Dollar and Its Rivals: Recession, Inflation and International Finance.* London: New Left Books, 1981.

Parker, M., and J. Slaughter. *Choosing Sides: Unions and the Team Concept.* Boston: South End Press, 1988.

Pentland, H.C. *A Study of the Changing Social, Economic, and Political Background of the Canadian Industrial Relations System.* Draft Study prepared for the Task Force on Labour Relations, 1968. Mimeo.

Piore, M., 'Perspectives on Labor Market Flexibility.' *Industrial Relations* 25, no. 2 (1986): 146–66.

Piore, M., and C. Sabel. *The Second Industrial Divide.* New York: Basic Books, 1984.

Pitsula, J., and K. Rasmussen. *Privatizing a Province: The New Right in Saskatchewan.* Vancouver: New Star Books, 1990.

Polanyi, K. *The Great Transformation.* Boston: Beacon Press, 1957.

Pollert, A. 'The "Flexible Firm": Fixation or Fact?' *Work, Employment, and Society* 2 no. 3 (1988): 281–316.

– 'Dismantling Flexibility.' *Capital and Class* 34 (1988): 42–75.

Potash Corporation of Saskatchewan. *Annual Report* (various years).

Poulantzas, N. *State, Power, Socialism.* London: Verso, 1980.

Price, J. 'Lean Production at Suzuki and Toyota: A Historical Perspective.' *Studies in Political Economy* 45 (1994): 66–99.

Rankin, T. *New Forms of Work Organization.* Toronto: University of Toronto Press, 1990.

Reich, R. *The Work of Nations.* New York: Vintage Books, 1991.

Reiter, E. *Making Fast Food.* Montreal: McGill-Queen's University Press, 1991.

Resnick, P. 'From Semiperiphery to Perimeter of the Core: Canada's Place in the Capitalist World Economy.' *Review* 12, no. 2 (1989): 263–97.

Richards, J., and L. Pratt. *Prairie Capitalism: Power and Influence in the New West.* Toronto: McClelland and Stewart, 1979.

Rinehart, J. 'Appropriating Worker's Knowledge: Quality Control Circles at a General Motors Plant.' *Studies in Political Economy* 14 (1984): 75–97.

– 'Improving the Quality of Work through Job Redesign.' *Canadian Review of Sociology and Anthropology* 23, no. 4 (1986): 507–30.

Roberston, D., et al. 'Team Concept and Kaizen: Japanese Production Management in a Unionized Canadian Auto Plant.' *Studies in Political Economy* 39 (1992): 77–107.

Russell, B. *Back to Work? Labour, State and Industrial Relations in Canada.* Scarborough: Nelson, 1990.

– 'The Subtle Labour Process and the Great Skill Debate.' *Canadian Journal of Sociology* 20, no. 3 (1995): 359–85.

– 'Labour's Magna Carta? Wagnerism in Canada at Fifty.' In C. Gonick et al. (eds.), *Labour Pains, Labour Gains: 50 Years of PC1003*, 177–92. Halifax: Fernwood, 1995.

– 'Rival Paradigms at Work: Work Reorganization and Labour Force Impacts in a Staple Industry.' *Canadian Review of Sociology and Anthropology* 34, no. 1 (1997): 25–52.

Saskatchewan. Bureau of Statistics. 'Saskatchewan Economic Statistics.' Regina, 1994. Mimeo.

– *Government Resource Series, Uranium*. Regina: Saskatchewan Education, 1990.

– *Revised Statutes of Saskatchewan*. Regina: Queen's Printer, 1978.

Sass, R. 'The Work Environment Board and the Limits of Social Democracy.' In J. Harding (ed.), *Social Policy and Social Justice: The NDP Government in Saskatchewan during the Blakeney Years*, 53–83. Waterloo: Wilfrid Laurier University Press, 1995.

Scott, J. *Domination and the Arts of Resistance: Hidden Transcripts*. New Haven: Yale University Press.

Shaiken, H. *Work Transformed: Automation and Labor in the Computer Age*. New York: Rinehart and Winston, 1984.

Shaiken, H., et al. 'The Work Process under More Flexible Production.' *Industrial Relations* 25, no. 2 (1986): 167–83.

Stepan-Norris, J., and M. Zeitlin, '"Red" Unions and "Bougeois" Contracts.' *American Journal of Sociology* 96, no. 5 (1991): 1151–1200.

Storey, R. 'The Struggle for Job Ownership in the Canadian Steel Industry.' *Labour/Le Travail*, 34 (1994): 75–106.

Swartz, D. 'New Forms of Worker Participation.' *Studies in Political Economy* 5, (1981): 55–78.

Taylor, L., and P. Walton. 'Industrial Sabotage: Motives and Meanings.' In S. Cohen (ed.), *Images of Deviance*, 219–45. Harmondsworth: Penguin Books, 1971.

Therborn, G. *Why Some People Are More Unemployed Than Others*. London: Verso, 1986.

Tomlins, C. *The State and the Unions*. Cambridge: Cambridge University Press, 1985.

Trist, E., and K. Bamforth. 'Some Social and Psychological Consequences of the Long-wall Method of Coal Getting.' *Human Relations* 4, no. 1 (1951): 3–38.

Trist, E., G. Higgin, H. Murray, and A. Pollock. *Organizational Choice*. London: Tavistock, 1963.

Trist, E., G. Susman, and G. Brown. 'An Experiment in Autonomous Working in an American Underground Coal Mine.' *Human Relations* 30, no. 3 (1977): 201–36.

United Steel Workers of America. 'Steelworker Guidelines for Participation in Work Reorganization.' Conference, 1992. Mimeo.

Wannell, T. 'Firm Size and Employment: Recent Canadian Trends.' *Canadian Economic Observer* (1992): 4.1–4.20.

Warrian, P. 'Labour Is Not a Commodity.' PhD dissertation, University of Waterloo, 1986.

Watkins, M. 'A Staple Theory of Economic Growth.' In W.T. Easterbrook and

M. Watkins (eds.), *Approaches to Canadian Economic History*, 49–73. Toronto: McClelland and Stewart, 1967.

– 'The Political Economy of Growth.' In W. Clement and G. Williams (eds.), *The New Canadian Political Economy*, 16–35. Montreal: McGill-Queen's University Press, 1989.

Wells, D. *Empty Promises: Quality of Work Life Programs and the Labor Movement.* New York: Monthly Review Press, 1987.

Williams, K., et al. 'Against Lean Production.' *Economy and Society* 21, no. 3 (1992): 321–54.

Witte, J. *Democracy, Authority and Alienation in Work.* Chicago: University of Chicago Press, 1980.

Wolfe, D. 'Technology and Trade.' In D. Drache (ed.), *Getting on Track: Social Democratic Strategies for Ontario*, 17–31. Montreal: McGill-Queen's University Press, 1992.

Wolpe, H. 'Capitalism and Cheap Labour-Power in South Africa.' *Economy and Society* 1, no. 4 (1972): 425–56.

Womack, J., et al. *The Machine That Changed the World.* New York: Rawson Associates, 1990.

Wood, S. 'The Transformation of Work?' In S. Wood (ed.), *The Transformation of Work?*, 1–43. London: Unwin Hyman, 1989.

Woodward, J. *Industrial Organization: Theory and Practice.* London: Oxford University Press, 1965.

Yarrow, M. 'The Labor Process in Coal Mining.' In A. Zimbalist (ed.), *Case Studies on the Labor Process*, 170–92. New York: Monthly Review Press, 1979.

Zuboff, S. *In the Age of the Smart Machine.* New York: Basic Books, 1988.

Index

Aboriginal women, working in mines, 44
Aboriginal workers: in potash industry,
    41; in uranium industry, 22, 44, 45,
    48–9, 71, 113, 182, 212n36
absenteeism, 68, 124, 134, 142, 169–70
Adler, P., and B. Borys, 225n31
affirmative action, in hiring, 22, 43, 44,
    48, 49, 211n28
Agrium Ltd., 23, 25, 53, 54, 62, 145, 150,
    155, 157; expansion of production,
    196; history, 87–8, 116, 208n98; prod-
    uct markets, 88; profits, 196
alienation, 125, 126, 130, 134, 165
American Department of Commerce. *See*
    United States
Amok Uranium Ltd., 37
arbitration, 163–4, 170; decision at Pot-
    ash Corporation of Saskatchewan–
    Cory, 6, 8
Arcadian Corporation, 9, 55
automation, 11, 21, 78, 111, 129, 132,
    155; in milling, 41, 120–2; in mining,
    84, 118–19; and post-industrialism, 18,
    125, 127–8, 141, 151–3

Babbage, C., 130
Bell, D., 129

benchmarking, 80, 98–9
Berggren, C., 138–9
Blauner, R., 126–7
Braverman, H., 11, 14, 130–3, 137,
    225n39, 226n41, 226n45
Bretton Woods, 13–14
bridges. *See* conveyor belts
Bright, J., 132
bumping rights, 60, 106
Burawoy, M., 71–2, 132, 138, 163
bureaucracy: in employment relations,
    12, 16, 90, 164

Cameco Corporation, 23, 26, 53, 54,
    65–6, 157; labour recruitment, 48–9;
    origins, 22, 37, 65–6, 208n96; priva-
    tization, 65, 66, 73, 74; profits, 65;
    share of production, 65
capital: mobility, 8, 10, 138, 198; turn-
    over time, 21, 22
Central Canada Potash Ltd., 53–4, 88, 89,
    109–10, 157; employment levels, 78;
    history, 75–6
Clement, W., 133, 213n19, 215n39,
    227n61
Clement, W., and J. Myles, 129
climate surveys, 25–6; at Agrium, 92; at

Central Canada Potash, 76, 79, 82, 84, 87

coal mining, 133–4

co-determination, at Central Canada Potash, 77, 82

collective bargaining: and Fordism, 12, 13, 17; structure of, 20, 163–4

Cominco Fertilizers. *See* Agrium Ltd.

Cominco Ltd., 88

company cultures. *See* cultures of employment

competency development. *See* training, Agrium

competition, international, 14, 100

contingent work. *See* lay-offs; seasonal employment

continuous flow production, 11, 111, 112, 118, 121, 125–6, 130, 131, 137, 140, 149, 157, 161

continuous improvement, 16, 17, 19, 23, 25, 89–90, 92–3, 95–103, 107–8, 110, 137, 139, 145, 169, 173, 194, 195, 221n93; criticisms of, 139–40, 166; union responses to, 104–5, 106, 107–8, 175, 222n115, 222n119; and workplace injuries, 189

continuous mining machines, 116–17, 118, 133

contracting out, 20, 28, 63, 69, 82, 184

conveyor belts: underground mining, 117–18

cooperative wage study (CWS), 60, 132, 165, 179; at Agrium, 90–1, 106, 220n65; at Cameco, 67; at Central Canada Potash, 85–6, 110; and grievances, 172–3; at Potash Corp. of Saskatchewan, 57, 59–62, 64

Corman, J., 55–6, 140, 213n14

corporate acquisitions, Potash Corp. of Saskatchewan, 9

corporate profitability, Potash Corp. of Saskatchewan, 9

craft labour, 125, 126, 128, 130, 131, 133, 136, 149, 155

crown equity participation program, 36

cultures of employment, 20, 21, 28–30, 53, 103, 108

customer satisfaction, 98–9, 102

cycle times, for jobs, 98, 139

deaths, at work, 184

democracy. *See* worker control

deregulation, of labour markets, 14

deskilling, 12, 15, 87, 124, 127–34, 137, 139, 141, 149, 151, 161

discipline, 27, 72, 162–4, 168–9, 171, 174, 177, 178, 182, 191; at Cameco, 68, 70

dissatisfaction, job, 124, 130

diversification. *See* vertical integration

Dohse, K., 138, 227n71

downsizing. *See* lay-offs

Duval potash mine, 31

economic rents. *See* royalties

educational background: potash workforce, 43; uranium workforce, 49

Edwards, P., and H. Scullion, 176

Edwards, R., 214n33

Eldorado Nuclear, 22, 36, 37, 65, 210n12

employee involvement programs, 25, 145, 166–7, 218n31; at Agrium, 90, 92, 193; at Central Canada Potash, 76, 77–8, 83, 84, 110

employee performance reviews: at Central Canada Potash, 76

employee recognition programs: at Agrium, 90, 92, 97, 100, 101; at Central Canada Potash, 84, 87, 218n36 employees' association, in potash industry, 7

employment equity. *See* affirmative action

employment stability. *See* job security

empowerment. *See* worker control

enterprise unionism, 21

Europe, potash companies in Canada, 54

European Economic Community, as potash export market, 9

exposure levels, attitudes towards, 48

familiarization periods, for job, 129, 150, 153, 156, 160

fast food workers, 11

First Nations workers. *See* Aboriginal workers

flexibility: and cooperative wage study (CWS), 61, 85; flexible accumulation, 22, 129; flexible specialization (flexibility theory), 18–19, 65, 206n77; and post-Fordism, 15–16, 17, 18, 85–6, 107, 139, 142, 143, 146; and post-industrialism, 127, 128, 131, 155–6, 161; and skill, 161; and unsafe work, 189; in work assignments, 73, 91, 107, 124, 134, 142–5, 148–9, 165

flexibility theory. *See* flexibility; flexibility and post-industrialism; post-Fordism

Florida Favorite Fertilizers, 54

fly-in/fly-out. *See* radical spatialization

focus groups, at Central Canada Potash, 82, 84

Fordism, 12–13, 16, 52–3, 64, 65, 98, 108–10, 141, 164; crisis of, 14, 18–20; employment levels, 13, 14, 70; and industrial relations, 19, 71, 75, 124, 138; national and global variants, 13;

and trade unions, 15–16; and welfare state, 13, 14, 37, 65

Former Soviet Union: potash production, 30, 210n5; uranium exports, 37, 211n18

four-X program. *See* continuous improvement

Friedman, A., 135

gainsharing, 17, 20, 66, 74, 92, 95, 96, 101, 105, 107–8, 145, 195, 196, 222n121

Gaullie, D., 129, 152

gender, composition of potash industry, 41

Germany: potash companies in Canada, 31; potash industry, 9, 55; potash production, 30

globalization, definition, 10

gold production, Cameco, 66

Gramsci, A., 12, 52–3

grievances, filed, 16, 27, 64, 70, 169–74, 177, 178, 182, 230n24; procedures, 55, 108, 163, 164, 166, 170

Hall, A., 59, 134, 213n23, 214n28, 214n29, 215n39

harassment: sexual, 191–2; in the workplace, 27, 167, 178, 189–94

Harvard Negotiation Project. *See* joint problem solving

Harvey, D., 22

health and safety, 27, 59, 100, 118, 167, 178, 181, 182; representatives, 55

health problems: work-related, 186, 231n34

Hirschhorn, L., 127, 128

homophobia, 191

human resources steering committee. *See* joint steering committees

Inco Ltd., 133, 134
indigenous workers. *See* Aboriginal workers
individuation, of employment relationships, 20, 59, 64, 74
industrial democracy. *See* worker control
injuries: official statistics, 184–5; at work, 185–6, 189
innovation, 137
innovation teams, at Central Canada Potash, 85, 87
internal labour markets, 16, 44, 60, 81, 173
International Metals Corporation (IMC), 53, 76
Israel, potash production, 30

Japan: imports of uranium, 38; Japanese production relations, 65, 103, 112, 136, 138–9
job analysis. *See* job observation
job classifications, 16, 20, 57, 60, 67, 73, 86, 91, 113, 144, 165, 214n35; for Aboriginal and non-Aboriginal workers, 46
job control unionism, 12, 16, 17, 45, 60, 61, 64, 85, 164–5, 167, 173, 214n33
job evaluations, 16, 25, 59, 60, 61, 67, 132, 165
job expansion, 8, 9, 15, 61, 85, 86, 121, 132, 134, 137, 138–9, 142, 145–6, 148–9, 156–7, 158, 160–1, 165, 199
job observation, as part of loss control, 58–9
job rotation, 134, 136, 139, 142–3, 144, 148, 150, 154–5, 156, 161, 165
job security, 21, 63, 65, 92, 95, 99, 100, 103, 106, 107, 126, 176–7, 195
job sharing, 100
joint potash bargaining council, 7

joint problem solving: at Agrium, 95, 97, 99, 105, 195; and post-Fordism, 18
joint steering committees, 24, 165; at Agrium, 92–7, 99, 101, 108; at Central Canada Potash, 76, 77–8, 81, 82, 83
journey operators, 91, 107, 150, 220n67
journey trades, 91, 107
just-in-time delivery, 16
just-in-time workforce, 22

Kali und Salz AG Potash, 9, 55
Kenney, M., and R. Florida, 137, 139, 140
Key Lake uranium mine. *See* Cameco Corp.
Keynesianism, 14, 15, 18, 19

labour markets, 13, 26, 29, 30, 50, 135; and Fordism, 13–16, 18; potash industry, 42–3, 81; uranium industry, 43–4, 46, 48–9
labour process debate, 15, 16, 95, 112, 130, 151, 161
labour scarcities. *See* understaffing
Laux, J., and M. Molot, 56
lay-off avoidance: at Agrium, 96, 100, 106, 108; at Cameco, 69, 109, 124; at Potash Corp. of Saskatchewan, 56
lay-offs: at Agrium, 89, 103, 106; at Central Canada Potash, 78, 81, 82; effects on unions, 7, 20, 63, 64, 177, 194; permanent, 3, 41, 47, 121, 134, 149; at Potash Corp. of Saskatchewan, 56, 65, 109, 215n50; at PCS-Allan, 7, 54, 63, 64, 173, 177, 182, 193; at PCS-Cory, 3–5, 54, 63, 144, 146; recall notices, 8, 64; resulting from removal of business, 6; role of seniority, 4, 8, 64, 71, 89, 195
lean production, 19, 20, 86, 87, 136–7, 138, 146, 189

lines of progression. *See* job control
    unionism
loadout: employment levels, potash, 121;
    employment levels, uranium, 122;
    operators, potash, 121
long distance commuting. *See* radical
    spatialization
loss control, 57–9, 61, 62, 66, 109,
    214n28, 214n29

maintenance workers: potash, 42, 86; ura-
    nium, 48–9; working conditions, 111,
    112, 149–58, 160
management, right to manage, 6, 12, 59,
    77, 85, 92, 93, 170, 181, 198
management by stress, 19, 189
market determinism, critique, 28
Marx, K., 52, 137
masculine work cultures, 190
McBride, S., 19
mechanization, 21, 40, 111, 112, 118,
    132, 133, 134, 141, 149
merit pay, 102
migrant labour systems, 22, 71–2, 73
milling: potash, 120–1, 123, 224n10; ura-
    nium, 122–3
mill workers: potash, 42, 121, 127, 142,
    143; uranium, employment levels,
    122–4; working conditions, 149–58,
    160, 229n102
mine borers. *See* continuous mining
    machines
mine operators: potash, 42, 86, 117–18,
    127, 142, 143; uranium, 114; working
    conditions, 149–58, 160
mining departments. *See* occupational
    groups
mission statements: at Agrium, 90; at
    Central Canada Potash, 79, 82, 87
mistrust, 90, 103, 194

monitoring of workers, 64, 68, 72, 74, 89,
    114, 119
more with less, as theoretical theme, 8,
    10, 100, 140, 141, 148, 160–1, 173,
    189, 199
multiskilling. *See* reskilling
multitasking. *See* job expansion

National Labor Relations Act. *See* Wag-
    ner Act
nationalization, of Saskatchewan potash
    industry, 31–2, 33–4, 54, 55
New Democratic Party/CCF: and potash,
    30, 31, 34, 56; and uranium, 36
needs assessment. *See* training, Central
    Canada Potash
neocorporatism, 13, 17
New Brunswick: potash mines, 40; share
    of potash production, 30
ninety-day notice. *See* technological
    change, labour legislation pertaining
    to.
nitrogen: production, 9, 55, 88; use in fer-
    tilizers, 30
Noble, D., 130, 225n39
non-union labour, 20
Noranda Inc., 75, 79, 81, 88
northern residential status, in hiring deci-
    sions, 44

obtaining employment. *See* labour mar-
    kets
occupational backgrounds of parents:
    potash workers, 41, 50; uranium work-
    ers, 46–7, 50
occupational groups, at the mines, 24–5,
    113, 114
Ontario, uranium mines, 36
Ontario Hydro, 37, 211n19
outsourcing. *See* contracting out

overcapacity. *See* potash, inventories
overtime work, 63–4, 89, 144, 193–4, 198; and grievances over allocation, 172–3

pace of work, 146, 153, 156, 179
Panitch, L., and D. Swartz, 19–20
parallel apprenticeships. *See* training, Central Canada Potash
Parker, M., and J. Slaughter, 139
participation: in local unions, 174–5
paternalism, employer, 67, 71–3, 126
pattern bargaining. *See* collective bargaining
payment-for-knowledge, 17, 20, 73, 136, 195
peer review, 25, 136, 166; at Agrium, 102; at Central Canada Potash, 84
Pennzoil Corporation, 31
Pentland, H.C., 162
performance sharing. *See* profit sharing
phosphate: production, 9, 55, 76; use in fertilizer, 30
piece rates, 176
Piore, M., and C. Sabel, 18, 206n77
Polanyi, K., 28
political apparatuses of production, 11, 15, 27, 53, 161, 162–4
Pollert, A., 19
post-Fordism, 11, 26–7, 82, 84, 89, 108–10, 136–7, 198–9; advocates of, 164, 166, 168; and conflict, 171, 174, 193–4, 196–7; critiques, 19–20, 80, 140, 166–7; industrial relations, 17–18, 74; and unsafe work, 188; and welfare reform, 18
post-industrialism, 10–11, 18, 112, 124, 127, 129, 131, 137, 140, 141, 149; critiques, 128, 130; and the labour process, 121; and skill, 150–3, 155–6, 158–6

potash: Canadian exports, 30, 33, 35; Canadian production levels, 30; crisis in industry, 7, 35; employment levels, 5, 40, 116, 117; geology, 114–15, 117, 119; inventories, 4, 35, 50, 62, 63, 196; new mines and mills, 4; obtaining employment, 42; origins of Canadian industry, 30; prices, 4, 31, 33–6, 51; production quotas, 5; product lines, 4, 5–6, 7, 63, 120–1; profitability, 35–6, 51; uses of, 30; value of production, 39
Potash Corporation of America (PCA), 30; mining divisions, 9; sale to Potash Corp. of Saskatchewan, 9, 54
Potash Corporation of Saskatchewan (PCS), 53, 54, 145, 157; divisional structure, 4, 7, 31; globalization of, 10, 55; market share, 9, 55, 62; origins, 54; PCS-Lanigan, 4, 7; privatization, 9, 36, 54; production capacity, 54–5; profitability, 9, 63, 201n5, 215n45
potash mills: employment levels, 3, 121, 123
potash resources act, 5
Price, J., 138
privatization: Cameco, 37, 65–6, 73, 74; Potash Corp. of Saskatchewan, 9, 36, 54
production norms. *See* production targets
production politics, 53, 65, 110, 161, 162–4, 177, 187
production targets, 146–7, 148, 153–4, 156, 158, 179, 224n6; in potash, 119
profit sharing, 101; at Central Canada Potash, 76, 79, 80–1, 82, 84, 85, 87, 110, 177, 218n20
protest demonstrations, 4, 64

Proudfoot study. *See* scientific management, Agrium

quality circles, 96, 140
quality of working life, 96, 135–6, 218n31
quit rates. *See* turnover rates

race: and division of labour at Cameco, 44–5, 70
racism, 191–2
radically marketized industrial relations, 27, 56, 63, 64, 78, 138, 145, 173
radical spatialization, 21, 22, 27, 44, 48, 49, 68, 69, 70–1, 72, 73, 138, 170, 173, 193
Rankin, T., 136, 228n79, 230n14
Reaganomics, 14
reassignments, work tasks. *See* flexibility, in work assignments
red product. *See* potash, product lines
refining. *See* milling
regulation school, 18
residential facilities, at Cameco mine, 69
resistance, 89, 132, 135, 168, 170, 175, 178
reskilling, 15, 18, 19, 20, 49, 73–4, 81–2, 85, 102, 126-9, 132–3, 134, 135, 136–8, 139, 140–1, 142, 143–4, 148–52, 156, 158–61, 165, 199, 228n74, 228n79, 229n102
responsibility, 126–7, 133–6, 139, 153, 156, 165
responsible autonomy. *See* responsibility
restructuring: definition, 10; economic and industrial, 3, 14
return-to-work bonuses, 64
Richards, J., and L. Pratt, 210n9
risk management. *See* loss control
Robertson, D., 139

robots, use in mining, 118–19, 223n1
rosters, for hours of work, 69, 71
royalties, potash industry, 30–1

sabotage, 64, 119, 134, 175–8, 230n30
Saskatchewan: potash mines, 40; share of potash production, 30; uranium mines, 36
Saskatchewan Mining and Development Corporation, 36, 37, 65
Saskterra Potash Corp. 9, 54
Sass, R., 56, 213n14
scientific management, 11, 12, 14–15, 16, 18, 19, 24, 95, 124, 125, 127–31, 133–4, 135, 136, 137, 138–9, 153, 164–5; at Agrium, 88–9, 90, 91, 103, 104, 119, 173; at Potash Corp. of Saskatchewan, 59
seasonal employment, at Potash Corp. of Saskatchewan, 54, 63, 64, 172
second incomes, 197
second jobs, 49–50, 70, 198
seniority: for Aboriginal and non-Aboriginal workers, 46; at Agrium, 90, 106; at Cameco, 44; and dual labour markets, 45, 71; and grievances, 172–3; in potash industry, 41; role of, 16, 17, 20, 41, 164; and training, 81
service sector, 11, 41, 42, 46, 47, 48, 129, 131, 140; Agrium as a service company, 92, 99; and second jobs, 49–50
severance packages, 4, 8, 100, 106, 202n20
sexism, 191
Shaiken, H., 130, 225–6n39
Shell Oil Co., 218n27, 230n14
shortcuts, in work practices, 176, 187
skill, 16, 17, 18, 27, 50, 60, 76, 99, 101–3, 110, 112, 125–7, 128–9, 131, 132,

134, 141, 149–50, 152, 156, 158, 175,
225n38, 226n44, 226n45
skill enhancement. *See* reskilling
social division of labour, 10
Societe National Elf Aquitaine S.A., 55
socio-technical approaches, 136, 140,
203n36
South Africa: potash companies in Canada, 31, 54; uranium exports, 37
southern Africa, mining industry, 22, 72
space, as a power resource, 67, 70, 74
spatial mobility. *See* capital mobility
staples: products, 21, 29, 32, 35; theory,
29
state ownership. *See* nationalization
statistical process analysis, 98, 103, 166
steelworkers. *See* unions
Stelco Ltd., 140
stock purchase plans: at Cameco, 73,
74; at Central Canada Potash, 76, 81,
82, 87
strategic choice, 125, 128, 134, 135, 136,
141
strategic opportunism, 103–4, 222n111
stress, job, 20, 27, 139, 167, 178, 189,
199
strikes: at Agrium, 106, 107, 175, 177,
194, 195–7, 198; at Cameco, 70; at
Central Canada Potash, 76, 169,
217n6; at Potash Corp. of
Saskatchewan-Allan, 59, 60; PCS-
Lanigan, 7, 202n19
suggestion programs. *See* employee
involvement programs
supervision, levels of, 114, 133, 134,
146–7, 148, 160–1
supervisors, 86, 95–6, 100, 102, 114, 126,
166, 172, 173, 187, 192–3; selection
of, 179, 181–2
surface lease agreements, in uranium
industry, 22, 43, 44

survey design, of mining workforces,
24–5, 26, 44

Taiichi, O., 137, 138
tariffs, on potash, 4
Taylor, L., and P. Walton, 176
Taylorism. *See* scientific management
technical division of labour, 10, 17, 59,
61, 64, 130, 133, 134; in potash, 118;
in uranium milling, 124; in uranium
mining, 113–14
technological change: arbitration of, 6;
bargaining over, 5, 6; definition of, 5,
6; labour legislation pertaining to, 5;
and lay-offs, 7, 78, 85
technological determinism, 130
technology: control over, 152–3; control
over adoption, 179
Texasgulf Corporation, 9, 55
Thatcherism, 14
total quality management, 16, 66, 83, 92,
96, 136, 218n31
toyotism, 138
trade union act, Saskatchewan, 5, 6
training, 18, 129, 133, 139, 149–50, 152,
166, 179, 181; at Agrium, 90, 93, 96,
97, 98, 100, 101, 102, 107; at Cameco,
49, 73–4, 142, 216–17n80; at Central
Canada Potash, 76, 81–2, 85, 86–7,
218n23, 219n47; potash industry,
142–3
training modules, 24; at Potash Corp. of
Saskatchewan, 57–8, 59, 61, 62
travel time, to mine faces, 116
Trist, E., 134–5, 136, 227n71
turnover rates, labour, 44, 49, 64, 91, 142

understaffing, 146–7, 148, 157–8, 160
unfair labour practices: at Central Canada
Potash, 80; at Potash Corp. of Saskatchewan, 5, 7

unilateralism, employer, 20, 59, 67, 72, 76, 80, 105, 106, 108, 157, 166, 181
unions: certifications, 7, 15, 20; Communication, Energy and Paper Workers, 7, 198; contract ratification votes, 25; declining power, 20, 67, 167, 174; densities, 13, 17, 20; national–local relations, 80–1, 104–5, 106; relations between local branches, 7, 64, 196, 198, 231n6; solidarity, 71; strategies re lay-offs, 8, 198; United Steelworkers of America, 7, 23, 59, 198, 222n110. *See also* employees' association
union-management cooperation, 19, 20, 23, 72, 80, 95, 99, 105, 106
United States: American Department of Commerce, 4; as export market for potash, 4, 33; imports of uranium, 36, 37–8; potash companies in Canada, 31, 54; potash producers, 4; potash production, 30
unsafe work, 186–7; responses to, 187–9, 194
Uranerz Uranium Ltd., 37
uranium: Canadian production levels, 37; domestic sales, 37; employment levels, 40, 113; exports, 37; future mines, 37, 119; geology, 112–13, 119; inventories, 38; origins of Canadian industry, 36, 44; prices, 36, 37, 38, 51; profitability, 38, 51; uses of, 36; value of production, 39

vertical integration, at Potash Corp. of Saskatchewan, 9, 54–5
veto power, managerial, 77
Vigoro Corporation, 75–6, 217n2
visions and values. *See* mission statements
voluntary early retirement, 100, 106

wages: adequacy of, 50, 197; in potash, 3, 60
Wagner Act, 12, 15, 20
Wagner, R. *See* Wagner Act
Wells, D., 167
whipsawing, 20. *See also* post-Fordism, industrial relations
white option. *See* potash, product lines
Womack, J., 136–7, 228n74
women workers, 22; in potash, 41; in uranium, 44, 69
Wood, S., 138
Woodward, J., 125–6, 127, 128
work-effort bargain, 17, 23, 28, 50, 51, 64, 74, 101, 107, 125, 130, 140, 163
work environment, 178
work environment boards, 56, 213n13
worker control, 18, 20, 27, 136, 141, 164, 165–6, 167, 178–81, 194, 198, 199; at Central Canada Potash, 84; desired levels of, 181–2; at Potash Corp. of Saskatchewan, 56
worker participation, 20, 83, 90, 165–6. *See also* worker control
workfare, 18
work histories, previous: in potash industry, 42; in uranium industry, 47
working to rule, 176
work intensification, 19, 137, 138–9, 140
workloads. *See* production targets
work teams, 15, 17, 128, 134, 135, 136, 146, 156, 165, 167, 169; at Agrium, 90, 92–4, 96–7, 98, 99, 100, 101; at Cameco, 73, 74; at Central Canada Potash, 77, 82, 84, 110

yellow cake. *See* uranium

Zuboff, S., 127, 225n33, 226n44